C000059257

Poisoned Spring

Poisoned Spring

The EU and Water Privatisation

KARTIKA LIOTARD
and
STEVEN P. McGIFFEN

PLUTO PRESS
www.plutobooks.com

First published 2009 by Pluto Press
345 Archway Road, London N6 5AA and
175 Fifth Avenue, New York, NY 10010

www.plutobooks.com

Distributed in the United States of America exclusively by
Palgrave Macmillan, a division of St. Martin's Press LLC,
175 Fifth Avenue, New York, NY 10010

Copyright © Kartika Liotard and Steven P. McGiffen 2009

The right of Kartika Liotard and Steven P. McGiffen to be identified as
the authors of this work has been asserted by them in accordance with
the Copyright, Designs and Patents Act 1988.

British Library Cataloguing in Publication Data
A catalogue record for this book is available from the British Library

ISBN   978 0 7453 2789 1   Hardback
ISBN   978 0 7453 2788 4   Paperback

Library of Congress Cataloging in Publication Data applied for

This book is printed on paper suitable for recycling and made from
fully managed and sustained forest sources. Logging, pulping and
manufacturing processes are expected to conform to the environmental
standards of the country of origin. The paper may contain up to
70 per cent post-consumer waste.

10   9   8   7   6   5   4   3   2   1

Designed and produced for Pluto Press by
Chase Publishing Services Ltd, Sidmouth, England
Typeset from disk by Stanford DTP Services, Northampton, England
Printed and bound in the European Union by
CPI Antony Rowe, Chippenham and Eastbourne

CONTENTS

ACKNOWLEDGEMENTS

The authors have had the kind assistance of many individuals in preparing this book. We would like in particular to thank those people who gave up their usually highly-pressured time to allow us to interview them: Olivier Hoedeman of Corporate Europe Observatory; Hans Blokland, MEP for the Dutch political party the Christian Union; Carolina Falk, Assistant General Secretary of the United European Left/Nordic Green Left group of Euro MPs, who was responsible for coordinating her group's response to the Water Framework Directive during its course through the European Parliament; David Hall of the Public Service International Research Unit and indeed all of his colleagues. Without PSIRU's detailed, meticulous research, this book could not have been written in its present form. Thanks also to others we interviewed: Sergiy Moroz of WWF's European Policy Office; Pieter de Pous of the European Environmental Bureau; Paulus Jansen, Member of the Dutch national parliament and spokesman on water issues for the Socialist Party; and finally all of the people with whom we have chatted at the numerous conferences and other events which we have attended, many of whom have offered valuable advice and information. We would also like to thank colleagues who have been helpful, particularly Kartika Liotard's policy assistant Thomas Gijselaers, and Audrey Wang, who typed up the interviews and much other material.

ACRONYMS AND ABBREVIATIONS

ACP	Africa Caribbean Pacific
ACU	Alternative Conjunctive Use
AquaFed	International Federation of Private Water Operators
BAR	Basins at Risk
BizClim	Business Climate facility
BWR	Basic Water Requirement
CAP	Common Agricultural Policy
CCFD	Comité Catholique contre la Faim et pour le Développement
CEO	Corporate Europe Observatory
CIS	Common Implementation Strategy
DG-Competition	European Commission Directorate-General for Competition
DG-Dev	Directorate-General for Development
DG-Envi	Directorate-General for Environment
DG Markt	Directorate-General for Internal Market and Services
EBRD	European Bank for Reconstruction and Development
ECPUTW	Economic Commission for Europe's Convention on the Protection and Use of Transboundary Watercourses and International Lakes
EEA	European Environmental Agency
EEB	European Environmental Bureau
EPAs	Economic Partnership Agreements
EPWN	European Public Water Network
EUWI	European Union Water Initiative
GATS	General Agreement on Trade in Services
HDR	Human Development Report

IFIs	International Financial Institutions
IISD	International Institute for Sustainable Development
IMF	International Monetary Fund
IMPEL	Network for Environmental Inspection
IPCC	Intergovernmental Panel on Climate Change
ISPA	Instrument for Structural Policy for pre-Accession
l/p/d	Litres per day
MDG	Millennium Development Goal
NGO	Non-Governmental Organisation
OECD	Organisation for Economic Cooperation and Development
PPIAF	Public–Private Infrastructure Advisory Facility
PPP	Public–Private Partnerships
PSEEF	Private Sector Enabling Environment Facility
PSIRU	Public Services International Research Unit
PUP	Public–Public Partnership
RSPB	Royal Society for the Protection of Birds
RWA	Regional Water Authority
SAMI	South Australia Murray Irrigators
SIWI	Stockholm International Water Institute
SNHP	Spanish National Hydrological Plan
SOE	State Owned Enterprises
TNC	Transnational Corporation
UN-DESA	United Nations Department of Economic and Social Affairs
UNECE	UN Economic Commission for Europe
UNEP	United Nations Environmental Programme
UNICE	Union of Industrial and Employers' Confederations of Europe (known since 2007 as Business Europe)
WCD	World Commission on Dams
WEF	World Economic Forum
WFD	Water Framework Directive
WISE	Water Information System for Europe

WSSCC	Water Supply and Sanitation Collaborative Council
WTO	World Trade Organisation
WWIN	Waterland Water in the Netherlands

INTRODUCTION:
A DANGEROUS SYNERGY

'Every step in the water crisis is due to greed.'
Vandana Shiva

This book is the first independent attempt to provide, in lay terms, a critical survey of EU water policy, both internal and external. In the particular context in which we have written it, however, we believe it would be irresponsible and short-sighted simply to examine the EU's laws and practices and leave it at that, however critical our perspective may be. Our subject is much broader: our starting-point is, after all, the substance on which all known life depends. Both of the authors were brought up in well-watered countries – as they are respectively Dutch and English, this is to put it mildly – where a permanent, reliable supply has long been taken for granted.

This is still, despite periodic shortages and increasingly frequent extreme weather events, generally the case for those of us who live in the developed world. It is far from being the case, of course, in most developing countries. Yet it is not simply solidarity that moves our concern, for unless drastic changes of attitude and policy take place, it may not be true for very much longer in our rich, rainy, northern homelands either.

The world is facing a serious and deteriorating water crisis: the human race is running out of water. The quantity of water on and within our planet remains more or less stable. It is the amount available to each of us which is falling with overuse, geopolitical change, changes in climate and changes in technology. And we need so very much of it.

As well as to drink, keep ourselves clean and prepare our meals, we need it primarily to grow our food and for other necessities, to the extent that agriculture consumes more water than any other

1

human activity. We need it also for manufacture and to provide all kinds of services. Not only is a great deal of water used in agriculture and industry, moreover, but a huge quantity is polluted by these activities beyond safe use. We need it to produce energy, not only directly from hydropower – which in fact consumes very little – but in power stations (both coal- and oil-fired) and nuclear and geothermal facilities, all of which use huge amounts. We need it for travel and to move goods, for recreation and for its aesthetic and numinous qualities. Without drinkable water, we would of course die. With only sufficient to meet our very basic requirements, we could survive. But precisely what would be surviving is perhaps not anything we would want to become.[1]

This water crisis which is threatening to destroy us can be attributed to population growth and urbanisation, as well as to climate change. The number of people on the planet continues to grow, as does their concentration in greater numbers in fewer areas. The UN estimates that the amount of water available per head globally declined by a third between 1970 and 1990.[2] This book is not, however, another hand-wringing Malthusian tract, for population growth would be perfectly manageable within a global economy organised by freely cooperating sovereign states in order to meet people's needs. The causes of this crisis are to be found in economics, for it is economic forces which are driving those demographic processes. Paradoxically, these economic phenomena include both rising prosperity and persistent poverty, so that in common with almost all of the world's serious problems the crisis of water can be said to result at root from the ill-distribution of wealth.

In those parts of the world where significant numbers of people have become more prosperous, their water use has risen with their incomes. However, this is not simply about newly rich citizens of China or India enjoying their success, or the way in which this means that these countries' consumption per head has begun to rise towards levels found in the north. Far more serious is the concentration of wealth and power which is occurring as transnational corporations tighten their grip on the world's resources and thereby on its politics. The effects of this are many and malign, and amongst them is a guarantee of the continued mismanagement of water, for such mismanagement is inherent to corporate control, not only of water itself but of decision-making in general.

The neo-colonialism of the Green Revolution and of developments since also plays a central role in this sorry drama. The absurdity surrounding bio-fuels is merely the latest episode to be played out. It has been calculated that for the EU to meet its bio-fuel targets will cost the taxpayer €22 billion per year, while at the same time driving up food prices and generating environmental degradation on a huge scale.[3] As the world is constantly reshaped in the interests of the wealthy – countries, corporations and individuals – the spread of tourism and the sometimes massively inappropriate leisure facilities which go with it mean that we are faced with the spectacle of thirsty people looking out over their thirsty fields at well-watered golf courses and hotel swimming pools. Rivers and lakes run dry, groundwater hidden underground for millions of years moves closer to exhaustion each day and huge amounts of water are wasted, or used for the most trivial of purposes.

It is true that a warming world is bringing about changes in water availability which are unpredictable and often quite sudden, posing constant problems even for those decision-makers of good will, people who do their best to ensure that adequate quantities of water get to where they are needed. If Malthus has no influence on our thinking, however, neither does that other gloomy follower of a gloomy god, Jeremiah. This is not a book about a natural disaster. As the recent UN report 'Water for People, Water for Life' notes, this crisis can be overcome, not least because it is less about 'nature' than about 'water governance'. It is 'essentially caused by the ways in which we mismanage water'.[4] We should remember that with few exceptions scarcity is not an absolute condition but is rather, as Tunisian scientist and water activist Mohamed Larbi Bouguerra puts it, 'a social relation to things'. It is also a phenomenon capable of creating winners as well as losers, for 'capitalism can function only on the basis of scarcity'.[5] Problems created by human beings can be solved by them, and the authors remain optimistic that this problem both can and will be consigned to history.

It is often said that water is free and belongs to us all. This may be the case, but getting it to where it is needed costs money and requires decisions to be made about where that money is to come from. Water may be collected in the form of rain water, extracted directly from rivers and lakes, or drawn from groundwater sources such as aquifers. It may be recycled from waste water. Through the application of a range of technologies ancient and modern, it may be taken from the sea or from brackish groundwater sources and desalinised.

Having been obtained from one of these sources, it must then be distributed either to where it is most needed or to those most able to pay. This involves political, social and managerial decisions and, thus, questions of social justice and equity. These questions are the central subject of this book. As Bouguerra argues, the provision of water is the *sine qua non* of all other human social and economic activity and so occupies a special role in society and economy.

While water policy is duty-bound to ensure that it is economically viable, it must equally incorporate the need for social solidarity, cooperation with countries lacking water resources of their own, ecological responsibility and respect for the needs of future generations and other living beings that share the planet with us.[6]

Ensuring the extraction of sufficient water from the most appropriate available sources in the face of population growth and changes in economy, technology, politics and climate poses problems for the management of water for drinking, for sanitation, for agriculture and industry, for cooling systems and for a host of other purposes. As things stand, this management is failing, and, as a result, not only individual human beings but entire communities and their cultures are dying. It is not failing, moreover, because of human error or incompetence, or because the tasks involved are simply too great or too difficult. It is failing, instead, because of two closely-related phenomena: the subordination of social and environmental goals to the economic interests of powerful corporations; and the inability or refusal of the existing global ruling elite to develop systems of economic and environmental planning capable of underpinning our survival as a species. Vandana Shiva puts it more succinctly when she says, 'Every step in the water crisis is due to greed.'[7]

This subordination to greed and the accompanying failure of planning are both, to a large extent, the fault of the international institutions which have the resources and scope to enable things to be otherwise. This is true of the World Bank and other international financial institutions (IFIs), it is true of the World Trade Organisation (WTO) and it is true of the European Union.

The EU's malign effects would be bad enough if they merely concerned Europe. However, whereas we will be giving a mixed report of the EU's internal policies, we can see almost nothing of value in those which it applies beyond its boundaries. The EU, in common with the other international organisations listed above, now clearly exists to serve the wealthy, powerful countries and corporations which dominate its decision-making.

Climate Change

The relationship between climate change and water is complex and variable. In some parts of the world rainfall will increase, while others will suffer drought. There is complexity also of a different kind, for not only is it the case that water availability will be affected by the changing climate, but it is equally true to say that how we use and misuse water is a major factor in influencing that climate. As Shiney Varghese, a Senior Policy Analyst at the Institute for Agriculture and Trade Policy in Minnesota explains, 'the two sectors in the world that use the most water, chemical intensive agriculture and fossil-fuel based energy production, are also the biggest contributors to global warming'. Varghese cites a recent Greenpeace report which, she says, found that 'industrial, chemical intensive agriculture degrades the soil and destroys the resources that are critical to storing carbon, such as forests and other vegetation'. Added to this as a source of increased CO_2 in the atmosphere are the dependence of intensive agriculture on huge amounts of fossil-fuel-based chemical inputs and long-distance transport of foodstuffs as developing countries are encouraged to pursue export-based growth.[8]

Drought and Water Scarcity

While excess water can bring tragedy to individuals and communities, a lack of it means slow death for every living thing affected, including entire civilisations and whole species. All known complex life-forms depend on the ability of plants to produce energy from sunlight, a process for which water is essential. As rains fail to fall or natural or artificial water supply systems fail, plants go through a number of changes. Depending on the species, they may have strategies for dealing with prolonged water deprivation, but having exhausted these their metabolic processes will decline and fail. Plants deprived of water characteristically change in colour and droop, a result of the closure of their stomata, whereby the plant reduces transpiration, the loss of water through evaporation. In a healthy plant, evaporation produces what is called the transpiration stream: as water evaporates it is replaced as water is drawn upwards by tension, passing from the roots through the xylem, the plant's water-carrying vessels. If further water is not available, however, the plant will begin to wither and wilt. Depending on species, this process is for a time reversible, and the plant can make a full recovery. If the water deprivation continues, its growth will be interrupted, many plant species being able to suspend systems not directly connected to survival. Others, however, adopt a different strategy, producing seeds and thus guaranteeing the survival of their genes into another generation.[9] Each of these strategies has strengths and weaknesses, but each is limited in the duration of drought it will enable the plant, and ultimately its species, to survive.

Animals thus suffer both from the indirect effects of drought – the decline and disappearance of plant food sources used by themselves or by their prey – and from the direct effects of water deprivation. Both plants and animals may respond to a shortage of water by migration. Plants may do so through the dispersal of seeds, some of which may find their way to areas unaffected – or relatively unaffected – by drought. Animals, including our own species, will use their mobility as individuals to seek out places where water is more abundant.

In some cases, migration may not be necessary. If the drying out is an extremely gradual process, some plant and animal species as well as micro-organisms may undergo evolutionary change as selection pressures push them in a certain direction. All of these changes will profoundly affect the ecosystems of which each species forms a part. In turn, human societies, dependent on and shaped by (as well as shaping) these ecosystems, will attempt to adapt.[10]

Extreme Weather Events

Flood can be more selective and more survivable, but it destroys far more rapidly than does the most extreme drought. In very recent times, floods have affected not only developing countries such as Bangladesh and Vietnam, but many parts of Europe, while on the other side of the Atlantic the hurricane which all but destroyed New Orleans in August 2005 will not quickly be forgotten.

Floods destroy lives, livelihoods and property in an instant, hitting poorer communities particularly badly and their effects can be long-lasting, even permanent.

It is not only the increasing frequency of such events that has enhanced their significance, but the way in which humanity appears no longer able, despite our highly-developed climate science, to anticipate them. The UN estimates that at least 665,000 people died in 2,557 natural disasters in the last decade of the twentieth century alone, and that 90 per cent of these disasters were water-related. 97 per cent of the victims were from developing countries. The most important factor in these disasters was not the increasing frequency of extreme weather events, significant as that was, but a failure in planning. The effects, moreover, are cumulative, with, the UN says, 'both individual and state barely able to recover from one disaster before the next catastrophe strikes'. Although flood was the most often reported event, drought was responsible for a greater number of deaths.[11]

Bad Weather, Bad Politics, Catastrophic Economics

The history of such weather events demonstrates their importance in the provocation of major social upheavals and the creation and disappearance of cultures and civilisations. We believe that this is of immediate relevance, especially in the light of a global food crisis which is entirely a product of the unfortunate synergy of bad weather, bad politics and catastrophic economics. This is not limited to, but most certainly includes, the bad politics and catastrophic economics which have been applied to water supply. Water and its supply to everyone on the planet (which should pose no insuperable problems) is instead rapidly developing into a major source of international conflict. The market, touted as the invisible hand that will lead us all to paradise, is incapable of providing the means to address climate change or the problems associated with it, including the problems of water supply, sanitation and drought and flood.

Extreme weather events are increasing in frequency as a result of climate change.[12] The Intergovernmental Panel on Climate Change estimates that it is 'very likely' that 'heavy precipitation events' will increase in frequency over most areas, causing damage to crops, as well as soil erosion and waterlogging. The quality of groundwater will also be affected and water supplies will be contaminated. Mortality and morbidity will increase in affected areas. In addition, settlements, commerce and transport will be disrupted, pressures on infrastructure will increase and private and public property will be destroyed.

In the event of drought, while the results may be less acute they will possibly be even more damaging in the longer term. They will include land degradation, a lowering of yields, an increase in livestock deaths and a greater risk of forest fires. Supplies of food will also become shorter, leading to an increased risk of malnutrition and disease. Industry, agriculture, power generation and domestic and social life will all suffer, leading to increases in migration.[13]

The implications of these events for human societies depend, however, not only on their frequency and intensity but on the economic circumstances of the areas affected, as well as on political choices. The effects of climate change are comparable in many ways to the effects of war. Climate change of the scale and rapidity currently being experienced is a violent, disruptive and destructive process, a response to which demands careful decision-making, careful planning and meticulous implementation of the decisions and plans which result. Total war, in which entire populations are involved, has produced a response in modern times characterised by the socialisation of markets, a high degree of state intervention and planning, and the perception of the community as a single entity which must be nurtured and preserved. Such an approach in Britain and other combatant nations during the Second World War ensured a widespread acceptance of the need for burden-sharing and equitable treatment, as well as for the direction of economic and social activity towards a common goal.

It is potentially disastrous that climate change should become manifest at a time when 30 years of propaganda have led to the hegemony of concepts which reject and contradict any idea of solidarity or common purpose. To the fantasy of 'free markets' can be added an individualism and egoism which offer no possible response to the severe problems with which the world is faced. Given that top-down, dirigiste socialism has failed to solve the problems created by capitalism, the authors recognise that new and innovative forms of social ownership and economic planning will be needed. We remain, however, unashamedly in favour of the socialisation of essential services, the social control of production and of a high degree of economic planning to meet our needs and our legitimate, life-enhancing desires.

Climate change is beginning to have a tangible impact in an era when the huge commitment of public investment and endeavour which offer the only possibility of a mitigating response seem unlikely to be forthcoming. The European Union's promotion of private ownership of water supplies and its failure, internally, to deal with major problems of wastefulness, particularly the inefficient use of water in agriculture, makes it a cause of the problem rather than a body to which we can look for solutions. The dominant economic ethos in Brussels and almost every EU capital is neo-liberalism, an orthodoxy embodied in every treaty since Maastricht. This has been dented by the financial crisis for which neo-liberal economics is entirely responsible, but hegemony is not so easily dislodged. Having no evidence to back it in the first place, more than evidence will be required to dislodge neo-liberalism from its throne. Its spread at a time of rapid climate change has created a dangerous synergy which threatens to undermine global social and economic stability, leading to ecological catastrophe, disorder, epidemic and eventually possibly even war.

The Problem of Uncertainty

Dealing as we are with the highly complex systems which produce the weather, precise forecasting of the future of the climate is impossible. As Vicky Pope, a senior scientist at Britain's Meteorological Office explains, 'scientists cannot give precise predictions of what will happen in 100 years, or even in the next 10 years.... However, we are able to give a range of possible outcomes of the world's actions and to assign probabilities to these.' Because of this, she continues, 'climate projections should be treated as an assessment of risk'. This does not mean, however, that these predictions should not lead to action:

> You wouldn't drive a car if you knew you had a 10% chance of causing an accident. Yet we continue to increase emissions, despite the fact that even if we could stabilise greenhouse gases at or close to today's levels there would still be only a(n) 80% chance of keeping global temperature rises somewhere between 2 and 3°C above pre-industrial levels. If we carry on increasing emissions at present rates then global average temperature rise will be 2°C by the middle of this century.[14]

In other words, what can be said for certain about rapid climate change is that, unless drastic action is taken, it will be catastrophic, even if the exact nature of the catastrophes which it will bring is subject to too many variables to permit precise prediction.[15]

However, some predictions can be made with more confidence of accuracy than this. While the likely extent of the problems associated with climate change may be difficult or impossible to gauge, the fact that such problems will occur and that they will be significant enough to demand an international, coordinated response is certain.

Recent research suggests, for example, that by 2050 five times as much land will be suffering under extreme drought as is currently the case, while 200 million people could be displaced. Professor Norman Myers of Oxford University has estimated that global warming could force 'as many as 200 million people' to become 'climate migrants' as their lives are 'overtaken by disruptions of monsoon systems and other rainfall regimes, by droughts of unprecedented severity and duration, and by sea-level rise and coastal flooding'.[16]

As Oli Brown of the International Institute for Sustainable Development states in a broad discussion of the issues associated with 'environmental refugees', while 'Professor Myers's estimate of 200 million climate migrants by 2050 has become the accepted figure', it is little more than guesswork, as 'nobody really knows with any certainty what climate change will mean for human population distribution'. If Myers' prediction is accurate 'it would mean that by 2050 one in every forty-five people in the world will have been displaced by climate change'.[17] Importantly, however, Brown though sceptical of precise predictions concludes by agreeing with Myers that 'the issue of environmental refugees promises to rank as one of the foremost human crises of our times'.[18] He cites with approval a UN official's view that:

> there are well-founded fears that the number of people fleeing untenable environmental conditions may grow exponentially as the world experiences the effects of climate change and other phenomena. This new category of 'refugee' needs to find a place in international agreements. We need to better anticipate support requirements, similar to those of people fleeing other unviable situations.[19]

Brown is typical of the great majority of scientists and those, including the authors, who have studied their research findings:

while quantified predictions are unreliable, if all reputable research points in the same direction, the question ceases to be 'what?' and becomes instead 'how much?' and 'when?'. While the speed at which our proverbial vehicle is travelling and the distance we must cover to get there may be hard to pin down, the handcart's destination is known to all.

Water, Health and Development

The problems related to water are matters of quality as well as quantity. Fully half of the world's hospital beds are occupied by someone suffering from a water-related illness.[20] These diseases have both direct causes in the quality of water available and indirect causes to do with quantity. In other words, people who do not have sufficient access to safe water supplies will be forced to drink, cook and bathe in unsafe water, and many of then will get sick as a result. More than a billion people worldwide have no access to safe drinking water and far more than twice that number have inadequate sanitation, while it has been estimated that half of the world's population is routinely exposed to polluted water.[21] The result is a global epidemic of parasitic and other water-borne infections, including recurrent and resurgent cholera and a host of other conditions from the mildly debilitating to the invariably deadly.[22] For working people in countries lacking basic social security systems, however, even mildly debilitating diseases can lead to catastrophic loss of income, creating a knock-on effect which can set in motion a downward spiral, dragging the victim and his or her dependants downwards into brutal and inescapable poverty. Yet all that would be needed to eradicate such problems is the application of simple technologies well-known in Victorian times and costing what are in truth risible sums of money. It costs far less to keep people alive than it does to kill them.

The European Union and the Politics of Water

To be fair to the European Commission, it is easy to find many of the above points made by its Directorate-General for Environment

(DG-Envi), or Directorate-General for Development (DG-Dev), and it cannot be said that the European Union has failed to show an awareness of the problems and of their urgency. Measures to ensure the quality of Europe's water resources constitute a major body of EU environmental legislation, covering a range of issues including groundwater conservation, drinking water quality, bathing water quality, chemical contamination, flooding, conservation of wetlands, river and marine resources and urban waste management.[23] However, the effectiveness of such legislation has been persistently undermined by the overexploitation of water in agriculture and in industry and by the polluting effects of much economic activity. According to the European Environment Agency, an official EU body responsible for providing expert advice to the Commission, 'Europe's groundwater is endangered in many ways. Significant pollution by nitrate, pesticides, heavy metals and hydrocarbons has been reported from many countries.'[24]

The EU Water Framework Directive (WFD) introduced in 2000, though still not fully implemented, seems at first sight a bold approach to the internal (to the EU) aspects of the problems involved. It sets out a list of targets and target dates, covering groundwater quality, drinking and bathing water, flood control and drought avoidance, waste management and pollution. However, progress towards the achievement of these goals has, in most areas of policy, been less than impressive.

This is, in some limited areas, inherent in the WFD itself. In far more cases it is a product of the corporate domination of EU decision-making. Evident since the Treaty of Rome was signed in 1957, growing stronger every year since, this domination has intensified massively since the turn of the century and now amounts to a virtual dictatorship of the corporations. The result is that potentially effective policies are undermined while those effective only at lining shareholders' pockets are promoted. Even though public ownership continues to dominate water supply in Europe and most of the rest of the world, utilities providing essential services are often forced to behave like private corporations. This is why so little has been done to counter the enormous waste which characterises the system of water supply, why water continues to be seen as a commodity and therefore a legitimate source of profit, why inappropriate crops continue to be grown in areas of inadequate rainfall and why there has been no examination of the broad sweep of EU policies to see where water could be conserved or supply improved.

While the WFD may arguably be inadequate to its task – whether it is or not is the subject of Chapter 6 – it is the EU's external policies which are far more obviously designed to serve the interests of European corporations, rather than those of the purported beneficiaries. Together with the IMF, World Bank and WTO, the EU has persistently promoted increased private sector participation in water and sanitation services. This fanatical devotion to the 'market' as a solution to the problems involved in organising water supply flies in the face of all available evidence. Measured by the proportion of the affected population gaining access to clean water, privatisation of water supply and deregulation of the market surrounding it have failed to record a single success. Public ownership does not guarantee that problems will be effectively addressed, but it is the *sine qua non* of a successful approach.

The European Union Water Initiative (EUWI), which governs much of the EU's approach to water policy in developing countries, is with almost touching transparency designed to increase the world market share of European corporations. Seen in this light, the Millennium Development Goal for water, which includes a commitment to halve the number of people in the world who do not have ready access to a clean and adequate water supply, becomes nothing more than a business proposition.

The problem goes beyond this, however, as a whole range of EU policies in effect deprive developing countries both of water itself and of the means to tackle water shortages. Internationally, water supply reflects global inequalities, and in the most grotesque fashion. Extravagant sporting and leisure facilities, many built with European capital by corporations featherbedded by EU policies, are built alongside barrios lacking any safe water supply. Agricultural regions in easy reach of airports are adapted to serve the tastes of rich consumers in the north, producing fruit, vegetables and flowers. The result, apart from ensuring that exotic flowers can be had on Valentine's Day and raspberries served with Christmas dinner, is to put further pressure on dwindling water supplies in Africa and Asia. The EU has the means to tackle these problems, but chooses instead to transform itself into Fortress Europe, as more and more people from societies squeezed dry of resources attempt to migrate into Europe in search of work.

A Call to Activism

Water policy, in its many aspects, reveals the real nature of neo-liberalism, the latest phase of the sequestration of the world's resources into private hands and their exploitation for profit. This process has been proceeding for hundreds of years. Since its inception as a dominant mode of production in English agriculture from the sixteenth century onwards, capitalism has followed up the legalised theft of English farmers' lands and livelihoods known as 'enclosure' with a series of further enclosures: of land on the global level, of time, of every means of livelihood and, through that, of the human will itself. What we are witnessing is what French theorist Paul Ariès calls 'hypercapitalism', a system based on the eradication of any kind of value except exchange value, a system which goes beyond commodification. Hypercapitalism rests on a hegemonic ideology which refuses to tolerate the persistence of any sector which renews itself financially by any method other than the straightforward financial transaction between 'customer' and provider. Nor can this ideology 'allow to be left fallow any unexploited source of profit'.[25] This is true not only because to do so would be to forgo an opportunity to accumulate wealth, but, in the longer view, because doing so would represent an admission that in some cases the market may not be a suitable instrument.

There is, neo-liberalism admits, certainly an area of life where market relations would not be appropriate. This is why the family and the concept of 'private life' have survived four centuries of capitalism's onslaught. There is simply no area of *public* life for which relationships determined by the market are unsuitable. To admit that there may be would be seen by the ideologues of this fantasy as the thin edge of the wedge. Looked at in this light, water is very far indeed from being anything special. Like health care, it is something which people most definitely need and therefore the ideal commodity, better even than tobacco or heroine, as the need is inherent and does not have to be created. As Ariès says, given this mentality, it would be naïve to imagine that anything previously regarded as the common property of humanity should be kept out of 'the line of fire'.[26]

The relationship between water, the environment in general and the economics and politics within which environmental phenomena are contextualised provides the opportunity both for an illustration of how this process has worked and is working and for a way of discussing how we can combat this global resource grab. Whilst it is important for each and every one of us to conserve water as individual consumers and small producers, to lay too much emphasis on this would be at best to miss the point. At worst it would offer a deliberate distraction from that point. Individual behavioural changes are perhaps most important as a constant reminder of the need, at the levels where authority is exercised – locally, nationally and internationally – for the kind of policies needed to avert catastrophe and save lives. Such policies can only be achieved, moreover, if the kind of radical thought and action necessary to win the war against climate change, against drought and flood, is taken into every area of political, economic and social thought, reordering priorities and attempting to create structures that would make such a reordering effective, undermining, reorganising or getting rid of those which prevent this from occurring.

Structure of the Book

This book is a call to activism, as only active and well-informed citizens capable of organising systematic campaigns against the existing power will prove effective in the global resistance so urgently needed. It will not, however, be mere rhetoric, but will provide also a handbook of the problem as well as some pointers as to where to look for more detailed information.

In Chapter 1 we will attempt to define drought and water scarcity and deal with their implications for human beings, our societies and the living world on which we depend. We will look at the effects of a failure of water supply on plants, animals, and the ecosystems, the natural and created systems of which each forms a part. And we will examine the economic and social consequences of drought in the context of these changes, and discuss the extent of the problems of inadequate water supply and sanitation.

Having given this broad introduction to the contours of one major aspect of the problem, in Chapter 2 we will do the same for that other extreme water event, flood, and again we will look at its effects on plants, animals, ecosystems and economies and societies. We will also sketch out the political, economic and practical similarities and differences between drought and flood and look briefly at the particular situation of the Netherlands as a country which lives under constant threat of inundation. The chapter will end by looking at the controversial business of flood control, how it has been managed to date and how it might be better approached in the future.

Chapter 3 will look at conflict and cooperation in the face of water-related problems and crises on the global, regional, national and local levels. Climate change is, as we have noted above, only one factor feeding into the current crisis, so in this chapter we will look at other elements of the broad background: population pressure, spreading prosperity alongside persistent and deepening poverty, urbanisation and the failure of political will and economic planning.

Chapter 4 will look more directly at climate change and its effects in the face of this failure. We will examine in some detail the conclusions of the Intergovernmental Panel on Climate Change and their implications for water supply, as well as more generally on human societies and the ecosystems on which they depend.

Chapter 5 will begin to present, analyse and criticise the European Union's input to these issues by looking at the European Commission's obsessive drive to privatise, liberalise and deregulate and how this affects the water supply and other water-related services. We will demonstrate how this obsession, based not on evidence of the superiority of privatised services but on pure ideology, serves the interests of huge, corrupt and inefficient water multinationals rather than those of the people of Europe.

Chapter 6 looks at the broad range of water-related policies pursued by the EU, which principally means those grouped under the Water Framework Directive. We will see how the positive aspects of this measure are being undermined by vested interest and the kind of short-term thinking inherent to parliamentary democracies in which power has been skewed towards those who control concentrations of capital.

Chapter 7 will look at the EU's external policies and the way in which water politics, and specifically the European Union Water Initiative and attempts to force privatisation on to developing countries, form part of a broad pattern of neo-colonialism.

In conclusion, Chapter 8 will consider the effectiveness of resistance and of mitigating measures in the face of the deadly synergy of climate change and neo-liberalism which is threatening all of our futures.

Throughout, we will explain why the evidence we have accumulated convinces us that the so-called 'free market' can offer no solutions to the pressing problems which form our subject matter.

1

DROUGHT AND DEPRIVATION

Defining Drought

> 'Drought...was not just about business, family or personal failure. It was
> like war. It was about despair, depression, powerlessness...'
>
> Michael McKernan, *The Red Marauder*[1]

Precisely when does a dry spell turn into a drought? The answer
rests on four broad definitions. These are complementary rather
than being in conflict with each other, identifying different types
of drought, though ones which will often occur simultaneously,
or in sequence, one provoking the other. The interplay of these
definitions also reveals the multiple nature of drought, which
is both a natural, meteorological event and a highly political
phenomenon.

Almost all droughts which afflict human societies are avoidable.
They result from the failure to solve a problem, a failure to marshal
resources in the service of human beings when and where they are
needed. This may be a result of incompetence, corruption, or bad
luck. However, as capitalism, which is now the global economic
system, cannot exist without structural shortages, there must be
a suspicion that some droughts in some places occur because
somewhere somebody powerful wants it that way. This suspicion
can only be reinforced by examining history, which teaches us
that he (and it always is a 'he') who controls the water supply
controls everything.

We will return in due course to the relationship between global
capitalism and drought, which is one of the major themes of this
book. For the time being we want to go into those four definitions
in more detail. The first is meteorological *drought*, when rainfall

or other forms of precipitation are significantly below the average for a particular region and therefore do not meet the expectations on which farmers and others base their planning. Drought is not merely an absolute shortage of water, but is the relationship between supply and demand, where the former falls short of the latter. So what may be a sufficiency or even an abundance in one region or in one society, will be a drought for another. This is particularly pertinent because, as we all know, the climate is changing. The economic system which, through its ownership of the means of production of knowledge, has convinced much of the world that it is the only route to 'freedom' and 'prosperity', is leading us, possibly irreversibly, towards global catastrophe. In the meantime, sudden and unpredictable changes in weather patterns are creating smaller catastrophes, many of them involving water, or a lack of it.

Meteorological drought may in turn bring about hydrological drought, where river flows are lower than the norm, and water levels in rivers, lakes and groundwater fall. Temporary (usually seasonal) bodies of water may not appear, or may disappear prematurely. Hydrological drought lags behind meteorological drought because deficits in precipitation must generally accumulate before they become significant causes of shortfalls in groundwater, surface water, stream-flow and soil moisture. Changes in land use, such as the removal of forest or the draining of wetlands – always, ostensibly, in pursuit of 'prosperity' through 'growth' – can have an influence, as can the damming of rivers and the creation of new water storage facilities such as reservoirs.[2]

These two forms of drought are likely to lead to a third, agricultural drought, in which soil moisture levels fall below what is usual and what the farmer will have planned for. Crop yields decline and people go hungry, or must pay more for their food, or must work harder, possibly bringing less favourable land into cultivation. Animals become thirsty or suffer hunger as their fodder crops also decline or become more expensive.

Finally environmental drought involves a combination of the above three forms of drought, with synergistic impacts on the landscape, on natural environmental features such as forests or wetlands, as well as an increase in problematic events such as dust-storms. Plants must adapt or die out, animals must migrate, if they can, or see their numbers go into decline. Other plants and animals more adapted to a restricted water supply may move in to take their place and whole ecosystems will be transformed as chains of consequences radiate from the central event of there simply being less water than before.

There are also socio-economic impacts resulting from an interaction of these different forms of drought with human communities, while in the other direction human activities may provoke drought, either through the effects of climate change on particular regions or simply through excessive water use, waste, deliberately withholding water in pursuit of politico-economic ends, or the failure of delivery systems.[3]

A general working definition of drought would encompass all of these aspects. One group of researchers examining the phenomenon in Spain's Jucar basin came up with 'a significant circumstantial decrease of the hydrologic resources, during a timeframe sufficiently prolonged, that affects an extensive area and that has adverse socio-economic consequences'.[4] An Australian researcher puts it more succinctly, 'Fundamentally, drought occurs when there is a mismatch between the water available and the demands of human activities.'[5]

However, important to our thesis is a further definition of drought which the Spanish team adds to the above list: operative drought. According to hydraulic engineers Andreu and Solera, this is a form of drought which differs from all of the others in that it 'does not depend only on natural processes'. Operative drought is likely to be a major contributor to the process of converting a meteorological drought into a socio-economic problem. As the Spanish researchers explain:

> [in] developed water systems, once the requirements of water for different uses have been identified, if the available water resulting from natural sources and from the management and operation of the system does not meet these requirements, then it could be called an operative drought,

in order...to stress the importance of the operation of the system in the presentation and characteristics of this type of drought.[6]

Operative drought can occur because of poor decision-making by uncoordinated individuals, local, regional or national authorities, through the profit-maximising activities of corporations, or through lack or failure of international cooperation. Just one example of the latter is Lake Chad, which has shrunk to a tenth of the size it was 40 years ago. This is partly due to unusually low rainfall levels, but also to the spread of irrigated agriculture and the lack of any effective cooperation between the lakeside states – Chad, Niger, Nigeria and Cameroon – which might have minimised the damage caused by this, and in particular by the dams that have made it possible.

The knock-on environmental effects – the spread of *typha australis* weed as canals have dried out in the intensifying drought, and the subsequent flourishing of the quelea bird – have had a negative impact on all lakeside countries. The quelea bird likes to nest in *typha australis* but eats *oryza sativa,* better known as rice, the very crop for which much of the irrigation was intended to provide water.

The Aral Sea provides an even more extreme and better known example of over-exploitation of lake waters leading to disaster. In this case, that disaster was caused by decisions taken by a unified state characterised by technocratic hubris and an absence of popular consultation or democratic input. Since the collapse of the Soviet Union, however, the failure of the newly-independent lakeside states to cooperate has meant that nothing has been done to remedy the situation.[7]

Meteorological drought is beyond any short-term human control, hydrological drought much less so, therefore there are things that we can do to alleviate agricultural and environmental drought and certainly much that can be done to relieve socio-economic drought. And those things can be summarised as measures to avoid operative drought, measures which are fully understood and achievable, as will be clear, we hope, by the time you have read this book.

How Much Water Do We Need?

The notion of operative drought leads to a difficult question and one which has produced many answers: the question of just how much water is needed. Often, however, the variety of answers turns out to reflect simply widely differing interpretations of the question. Are we speaking about absolutely fundamental survival requirements in an emergency? Or something more permanent? Clearly, need will be affected by climate, but is it legitimate to take cultural factors into account? What about individual differences? And are we speaking of how much water a person needs to be delivered to her home, or every drop of use for which that same person's life is responsible? In the latter case, virtually everything she uses will have required some water at some point in its production and this must be factored in.

One of the most thorough studies of how much water is actually needed was carried out by US environmentalist Peter H. Gleick, author of several biennial reports on the world's water resources. Water is needed for drinking, waste disposal and treatment, the production of food and manufactured goods, generating energy and so on. The amount used will therefore vary with the level of development of a society, but also with a number of factors independent of that level. Gleick distinguishes between water 'withdrawal' – 'the act of taking water from a source to convey it elsewhere for storage or use' – even if that water is returned to the source afterwards, and 'consumption' or 'consumptive use', which is 'the use of water in a manner that prevents its reuse'.

The most obvious use of water is drinking the stuff. The amount a body needs, not surprisingly, varies with climate and also with diet and the size and level of physical activity of the individual. According to Gleick, the minimum requirement in a temperate climate is on average three litres per day, in a hotter climate perhaps as much as five litres. Added to this is the water required for sanitation. Although, as Gleick notes, efficient sanitation which uses little or no water is possible:

> where economic factors are not a constraint, cultural and social preferences
> strongly lean towards water-based systems.... Accordingly, while effective
> disposal of human wastes can be accomplished with little or no water when
> necessary, a minimum of twenty litres per person per day is recommended...
> to account for the maximum benefits of combining waste disposal and
> related hygiene, and to permit for cultural and social preferences.

For bathing, Gleick admits that data on necessary minima are
scanty, but recommends a 'basic level of service' of 15 litres per
day. Food preparation is certainly going to depend on cultural
factors, but here the available data are rather more extensive
and Gleick notes that 'studies in both developed and developing
countries...suggest that an average of 10 to 29 litres per day
appears to satisfy most regional standards and that 10 l/p/d will
meet basic needs'. Gleick's conclusion is that the basic water
requirement (BWR), the amount of water required to meet basic
needs over a long term, should be 50 litres per person per day.
This is made up in the following way:

> Considering drinking water and sanitation needs only suggests that the
> amount of clean water required to maintain adequate human health
> is between two and 80 litres per day...The low end of this range is an
> absolute minimum and reflects survival only. The upper end reflects a
> more complete satisfaction of basic needs using water piped directly to
> the house and toilet.

On this basis Gleick comes up with a figure of 25 l/p/d. Adding
water for bathing and cooking raises the total range to between
27 and 100 l/p/d. Gleick considers 15 l/p/d to be the minimum
acceptable level for bathing, and 10 l/p/d for cooking, which
gives him a total – a BWR – of 50 litres. To put this in context, he
quotes an estimate for average domestic water use in arid regions
of 100 l/p/d.[8] In western Europe people use on average at least
four times Gleick's BWR, while in North America they use more
than ten times that figure.[9]

Despite Gleick's meticulously considered work, we are left with
no clear answer to the question posed at the beginning of this
section. We would argue that this is because no single definition

is possible, as many variables are present. In short, it depends on what you mean by 'need'. This brings us back to the problem of defining drought. If operative drought is the existence of a gap between demand and supply, we must have some idea not only of the levels of demand in a particular society, but also what would be reasonable for that society. Only by eliminating profligacy and waste can the latter figure be arrived at, which should make these the most pressing tasks for almost any developed country. The extent to which this message has sunk home in one group of such countries will be seen in Chapters 5 and 6, when we look at the internal policies of the European Union.

Population Growth and Drought

The fact that the world's population is continuing to grow, with some 80 million people being added each year, has obvious implications for the demand for water. Population growth since 1970 means that there is now one-third less water per human head. Although there is, averaged out over the globe, plenty of water to go round, its uneven regional distribution would create problems even if social and economic inequalities were eradicated. As these inequalities are perpetuated by equally glaring inequalities of power, however, they are in reality widening.

Thirty-one countries, most of which are in Africa or the Middle East, are officially classified as facing 'water stress or scarcity',[10] and 17 more are projected to join them by 2025. If these predictions prove accurate, around a third of the world's population will be affected by some degree of water shortage.[11] The UN estimates that half the population of developing countries lives in 'water poverty'.[12] And the bad news is that it is the countries with the most pressing problems of water scarcity which have, in general, the fastest-growing populations. Sub-Saharan Africa is projected to double its population by the middle of this century, with North Africa and the Middle East seeing an increase of almost two-thirds.[13]

There is, however, more to the shortage of water than can be learned by dividing available water by the number of human beings and deciding there are just too many of us. Malthusian prophets of doom rarely offer to be the first over the cliff. Policy makers would be better employed devising ways of enabling people to take care of themselves and their children than lecturing the rest of us about family size. The fact that there is, on the one hand, a strong correlation in the contemporary world between poverty, lack of education, absence or weakness of a social state and absence or weakness of respect for women, and on the other hand, high birth rates, does not indicate that fewer births would mean greater prosperity, but that greater prosperity invariably leads to fewer births. Allow people access to resources, end the continuing plunder that is US and EU foreign policy, empower poor women and men and give them space to organise, to live, to breathe and to care for and enjoy the company of the children they do have, and birth rates will fall.

Despite the scandalous level of deprivation associated with access to drinkable water, the situation with regard to sanitation is far worse. Nearly half of the population of developing countries has no access to decent, effective sanitation, a proportion which reaches two-thirds in sub-Saharan Africa and South Asia and six-sevenths in Ethiopia. Countries which have experienced rapid and unplanned urban growth have seen sewerage systems break down under the pressure. Pit latrines and other poorly-designed home-made facilities contaminate the water supply. And all of this costs lives, livelihoods and money. As the 2006 Human Development Report (HDR) explains, 'studies show that the method of disposing of excreta is one of the strongest determinants of child survival, with the arrival of effective sanitation reducing child mortality by around a third'. Think of this when you perform your own next evacuation: one in three children who dies is losing her life for want of a decent place to shit.[14]

Sanitation is so important to human beings that they are willing, when given the opportunity, to invest heavily in it, so that the arrival of prosperity is invariably accompanied by the music of flushing toilets. Prosperity, however, brings its own problems to the issue of water availability, problems for which the flushing toilet, sweet as its song may be, is a fitting symbol. Worse than the somewhat water-profligate WC, however, is what goes on at the other end of the digestive process. As people become wealthier, they begin to demand not only more diversity but, specifically, more meat.

A recent report by the Stockholm International Water Institute (SIWI) looked at the impact of changing food consumption patterns in the face of rising prosperity on the allocation and use of water for agriculture. As the report pointed out, 'Unlike water use in industry, the high proportion of consumptive use in agriculture,[15] means that this water is lost for re-use or re-circulation in society.' People typically begin, as their real incomes increase, to replace the cereal products which are almost everyone's staple food,[16] with sugar and vegetables. Meat is, however, now absurdly cheap in relation to its real costs of production in almost every part of the world, so a relatively small rise in wealth will enable a large rise in its consumption. Between 1961 and 2002 wheat consumption tripled in Asia but the consumption of meat rose seven-fold.[17] In the nineteenth century, and in many parts of the world long after, this was not the case, and as working people's real incomes rose their consumption of meat tended to go from non-existent or ceremonial (weddings and other major events) to hebdomadal ('Sunday dinner'). The low price of meat internationally now means that people rising in income on a comparable curve to that of the English working class in the late nineteenth century, or the Dutch labouring classes somewhat later, can now very quickly afford to eat meat every day, if not at every meal. This may be pleasant for them as individuals, but it is potentially disastrous for the water supply. As the SIWI report explains, between 500 and 4000 litres of water are evaporated in producing a single kilo of wheat. This sounds a lot, and the seemingly wide range is due to variations in climate, farming systems, the length of the growing season and variety of wheat. The water consumption figures for meat, however, put these figures into perspective: to produce a kilo of meat takes between 5000 and 20,000 litres of water. The top end of this figure exceeds the minimum needed for a single human being to sustain a decent life for a whole year.[18] If you measure the input in terms of energy, 1000 calories – something over a third of the calorific requirement of an average man and about half that of an average woman – takes half a cubic metre of water if you are growing plant-based food and eight times that amount for animal-based food.

For this reason, the report continues:

> The production of meat from animals fed on irrigated crops has a direct impact on water resources, much more so than if the meat is derived from grazing animals and animals fed on residues. Irrigation water, withdrawn from rivers or other water bodies and returned back (*sic*) to the atmosphere by crop consumptive use, will not be available for cities, industry or the environment.[19]

The conclusion of this moderately-toned report is that the probability is 'that if today's food production and environmental trends continue, crises will occur in many parts of the world', with bio-energy demands compounding the likelihood of this.[20]

Global inequalities, not 'over-population', are thus clearly responsible for water shortages. As the 2006 HDR states: 'In the early 21st century debates on water increasingly reflect a Malthusian diagnosis of the problem. Dire warnings have been posted pointing to the "gloomy arithmetic" of rising population and declining water availability.'

Yet is the world 'running out of water'?

> Not in any meaningful sense. But water insecurity does pose a threat to human development for a large – and growing – section of humanity. Competition, environmental stress and unpredictability of access to water as a productive resource are powerful drivers of water insecurity for a large proportion of the global population.[21]

These are products not of simple population pressure but of a profit-driven neo-colonialist system. This is easily demonstrated if we look at figures for water use. As Tunisian academic and water activist Mohamed Larbi Bouguerra notes in his recent book *Water Under Threat*, when 'a German consumes ten times more water per day than an Indian, and an Israeli settler six times more than a Palestinian', a rising population can hardly be seen as the main problem.[22] True, some differences in consumption are due to 'natural' causes, the availability of water as a natural resource within a particular country or region. This does not begin to explain the huge disparities, however, between water available to individual human beings, which is conditioned far more by infrastructure. And the extent to which a country or region has the necessary infrastructure to deliver wholesome water is dependent on skewed and loaded political and economic relations of power. In the interests of a system driven by profit in which any degree of economic planning is hindered by vested interest and the prioritisation of the short-term and the (rich) individual over the long-term and the social, the problem of water shortage is being blatantly ignored by those who control resources, by the owners of capital and the political decision-makers who dance to their tune. This is why the most rapidly-growing countries in the world are those which are already short of water and why the same is true of the most rapidly-growing regions of the United States.[23] This would not be a problem if this growth were made possible by greater efficiency of water use, but it is in fact fuelled by increasingly unsustainable consumption of water from rivers, lakes and underground aquifers.

Population growth is, nevertheless, undeniably a factor in increasing demand for water. Along with industrialisation, urbanisation, the spread of irrigation and the relentless pursuit of profit, it has led to the waters of many of the world's rivers – great and small – being subject to what is termed 'over-allocation'. Quite simply, more water is being taken from them than the natural processes which replenish them can sustain. Enormous rivers such as the Nile, which waters 130 million citizens of ten different countries, and the Yellow River in China, peter out before they reach the sea during dry periods of the year. In the United States, the reservoir that supplies water to the cities of Phoenix and Las Vegas, which is fed by a seriously stressed Colorado River, has been estimated as having a 50 per cent chance of running dry by 2021.[24]

Rivers are often replenished from groundwater aquifers, which are also commonly a direct source of water supply. In addition, it is possible through careful management to conserve both surface and underground water sources by using the former during wet seasons and the latter during the drier periods of the year, a practice known as Alternative Conjunctive Use (ACU).[25] Groundwater can also be recharged from surplus surface water, or by using treated waste water. The quality of such water is often improved by passing it through the subsoil, reducing the overall cost of recycling it.[26]

Overexploitation is unfortunately far more common than is careful management, however. As a result, groundwater levels, like those of rivers and lakes, have fallen in many parts of the world over recent decades. According to the IPCC, moreover, with rare exceptions climate change is at worst a minor accomplice in the problems associated with aquifers, the falling levels being almost always simply the result of pumping surpassing recharge rates which have remained substantially unchanged.[27]

Overexploitation has numerous consequences, none of them good. Water must be brought from deeper and deeper underground, and below a certain level the concentration of minerals makes it unusable. In coastal areas, the fresh water in an aquifer will in effect float on a denser layer of saltwater and overexploitation will lead to the contamination of the former by the latter. This is known as 'saltwater intrusion' and it is already a problem in various parts of the world, including the Mediterranean countries, Miami, the Indian subcontinent and South East Asia. Overexploitation can also cause land subsidence, where the geological materials forming the aquifer compact and vital water storage space is lost for good. Biologist and water campaigner Constance Hunt gives the example of California's Central Valley, where 'compaction of over-drafted aquifers has resulted in a loss of nearly 25 billion cubic metres of storage capacity – equal to more than 40 percent of the combined storage capacity of all of California's human-made surface reservoirs'.[28]

As with food, fuel and almost every other vital resource, the problem is not that there isn't enough to go round, but rather that the economic system which increasingly dominates the planet leads inexorably to ever greater inequalities. Conservation and distribution are the real problems, for we are a very long way from anything which could be seen, almost anywhere in the world where people actually live, as an absolute shortage of water.

Urbanisation and Drought

Water shortage is one of the more important of a range of reasons causing people to migrate from countryside which they have previously farmed to take their chance in the city. So it is ironic that the very problem which they are escaping is being exacerbated by the process of urbanisation to which their decision is adding fuel.

Much of this urbanisation is informal and unorganised, involving the growth of huge shanty towns on the outskirts of cities in the developing world. The majority of the world's biggest cities are now to be found in its poorest countries. Apart from New York and Tokyo, where urban growth has joined together previously separate settlements of some considerable size, the top ten are all in low-to-middle income countries: Mexico (Mexico City), Brazil (Sao Paolo), India (Mumbai, Delhi, Kolkata), China (Shanghai), Bangladesh (Dhaka) and Argentina (Buenos Aries).[29] The UN acknowledges 20 agglomerations with more than 10 million inhabitants, with only Los Angeles and Osaka and the aforementioned New York and Tokyo in wealthy countries. According to the UN projection, it is the urban areas of poor countries where the bulk of the world's population will soon live: 'Virtually all of the world's population growth will be absorbed by the urban areas of the less developed regions, whose population is projected to increase from 2.4 billion in 2007 to 5.3 billion in 2050.'[30]

Urbanisation has a complex range of implications for water supply, especially when it is unplanned, or poorly planned. Runoff rates may increase downstream of cities as a result of a greater area being covered with the hard surfaces characteristic of towns, transport systems and industrial premises; or they may decline due to greater levels of extraction as populations rise. Pollution will also increase downstream and in any nearby lakes. Deterioration of the water supply in a city under pressure from a burgeoning population represents a major blow to the quality of life, leading in many cases to other negative impacts. The poor will often end up paying far more for their water than the rich, and criminals will be attracted by the predatory possibilities afforded by a failure of public services to function.[31]

Pollution as a Source of Water Shortage

In creating water scarcity, pollution of water courses can have the same effects as meteorological drought. Depending on the pollutants involved, contaminated water may be unusable, of restricted use, or require expensive decontamination before it can be used. According to the UN, the most widespread pollutant is human waste, followed by industrial wastes and chemicals and agricultural pesticides and fertilisers. Half of the population of the developing world is 'exposed to polluted sources of water that increase disease incidence'. The pollutants involved include 'faecal colliforms, industrial organic substances, acidifying substances from mining aquifers and atmospheric emissions, heavy metals from industry, ammonia, nitrate and phosphate pollution from agriculture, pesticide residues... sediments from human-induced erosion to rivers, lakes and reservoirs and salinization'. Although developed countries may produce more pollution, and are far from unaffected by a consequent reduction in water availability, dangerous levels of exposure are more likely 'in developing countries where institutional and structural arrangements for the treatment of municipal, industrial and agricultural waste are poor'. And things are not improving. Far from it. 'Levels of suspended solids in rivers in Asia have risen by a factor of four over the last three decades' and they have 'three times as many bacteria from human waste as the global average' while some Asian rivers have 'twenty times more lead than that in the surface waters of OECD countries.'[32] Delocalisation of European and North American corporate producers to developing countries is accompanied by pressure on those countries not to enact effective anti-pollution legislation or, if enacted, not to enforce it.

The developed world has had a considerable measure of success in curbing pollution since policy-makers woke up to the extent of the problem during the 1950s. The approach generally taken, however, may be difficult for developing countries to emulate, at least until they reach a certain level of income. In wealthy countries, as Constance Hunt explains, generally the approach has been 'to set technology-based treatment standards that specified the treatment levels required before an effluent could legally be discharged into a waterway'. More recently, standards have been aimed at protecting specific bodies of water 'and must take into account the cumulative effect of multiple pollution sources, the sensitivity of local aquatic species, and regional hydrological conditions. Effluent limitations and treatment technologies are then specified to ensure that the ambient water standards are not transgressed.'[33] In addition, developing countries are interested in attracting the very companies which are looking to move production from the north precisely because of such standards. Whether through corruption or because they are genuinely desperate to generate employment in order to combat poverty in their countries, legislators in developing countries have been unable to withstand this pressure. The result is that many of the social and environmental crimes committed by corporations as Europe and North America grew economically are being repeated in countries anxious to emulate their prosperity.

Pollution control of the kind favoured in the prosperous north is expensive in the short term and can only be tackled successfully by countries with the wealth, as well as the scientific and engineering know-how, to create and maintain the necessary systems. It is clearly better not to produce the pollutants in the first place. This requires analysing any production process in industry or agriculture, as well as the domestic environment, to see where emissions of pollutants could be eliminated or reduced.[34] Indeed, one easy way to reduce pollution is not to use more of a chemical than is necessary.[35] This would require some fairly inexpensive training for developing countries, but it would hit the profits of TNCs who make their living from selling those chemicals, as well as, ironically, from end-of-pipe pollution controls. As technology and training require the cooperation of governments, and as corporate influence on government is not confined to poor countries, this is simply not happening to the extent clearly needed.

Tackling pollutants as they leave the production process or enter the water course is difficult when one is dealing with 'substances that enter a waterway in a diffuse manner'. These include agricultural runoff, runoff from industry and residential developments, and pollutants entering the water via atmospheric pollution. Methods of control include 'the use of silt fences in urban construction projects, the planting of filter strips next to streams in agricultural areas, and the construction of artificial wetlands to cleanse storm water runoff from roads and parking lots in urban settings'.[36]

The problem, however, is that by imposing environmental standards on the south, wealthy countries can also institute a form of protectionism. On the other hand, if a developing country can be convinced that it can afford to adopt such standards, it may begin to win markets from richer rivals. Urging Pakistan to accept higher environmental standards, a group of researchers from the Institute for Sustainable Development in Islamabad concluded that such standards would 'ensure efficiencies and economies within the firm'. Also, the quality controls built into internationally recognised standards of production would enable Pakistan's industries to win and retain export markets. Beyond these immediate interests are the general benefits of a cleaner environment. Finally, the researchers argued that the costs of mitigating pollution are in any case much lower than generally believed, so that return on investment would be faster. For example, 91 per cent of polluting emissions from the textile industry and 66 per cent from the tanning industry could be eliminated, for what the researchers insist is a relatively modest outlay.[37]

The fact remains, however, that environmental standards are far lower in the developing world than they are in wealthy countries and that this has a direct impact on the water supply. In China, the production of a tonne of steel takes between four and nine times the amount of water that is used to produce the same in the

United States, Germany or Japan, yet the resultant emissions are much more polluting. Relocation of manufacturing industry to the south is thus exacerbating the problem of water scarcity, both globally and in the countries to which the polluting industries are relocating.[38] The solution is not to deprive developing countries of much-needed economic activity, but to cooperate with them in improving laws and their enforcement. To do this, however, corporate power over the legislative process globally, and corporate influence within the WTO, must be greatly reduced.

Climate Change and Drought

Droughts, in common with other extreme weather events, have become increasingly frequent over the last 40 years, especially in tropical regions. The effects of climate change are mixed, but overall precipitation over land areas has fallen globally and temperatures have risen, both clearly contributory factors. The recent report from the IPCC, moreover, concluded that it was likely that this trend would continue. The worst-affected areas in Europe are likely to be Spain, Portugal, western France and the continental interior. In Africa, where, by 2025, 460 million people will be affected by water stress due in part to meteorological drought, the worst hit will be the countries of the Horn of Africa, Kenya, Burundi, Rwanda, Malawi, South Africa and Egypt. In Asia it will be northern parts of China, the Pakistani coast and arid interior plains, parts of north-eastern India, the Philippines, Indonesia and parts of Japan; and huge areas of Australia. Only the Americas are largely left out of these firm predictions, presenting a much more mixed picture which makes prediction hazardous. Just a few areas, most importantly the south-western United States and the Canadian prairies, seem certain to experience lower levels of precipitation.[39] These matters will be dealt with in more detail in Chapter 4.

The Environmental Effects of Drought

The environmental consequences of drought have clear implications for human populations, for the land, cultivated and uncultivated, on which we depend, and for the aquatic ecosystems of rivers, lakes and ponds. Erosion of soils and transformation and reduction of soil quality is an ongoing drought-related phenomenon affecting much of the world, while an increased incidence of fire is evident in dryer, hotter regions.[40] More narrowly region-specific hazards may also be of significance, such as the grasshopper infestations which typically accompany drought in some western areas of the United States.[41]

Drought first hinders the growth of plants and then, if it persists, kills them. And plants sit at the base of the food chain. Even an exclusively carnivorous creature depends ultimately on plants for its food, even if eating exclusively carnivorous prey, or micro-organisms, or fungi. This means that apart from their own direct need for water, animals, including ourselves, need water because plants need it.

Plants of different species will react in a variety of ways, and with varying adaptability, to a shortage of water. By closing their stomata, many species are able to slow transpiration, enabling them to survive dry spells, though at the cost of slowed or interrupted growth. Others react very differently, accelerating their seed-production process and thus enhancing their chances of passing on their genes to another generation. Should the dry spell persist, however, none of these strategies is likely to succeed to the extent that the plant continues to be a viable food source for animals and humans. Crop plants are in many cases particularly liable to damage through drought. In rice and soya, for example, the xylem, the structure which land plants use to transport water, is highly vulnerable. When re-watering occurs as a drought ends, the xylem's ability to conduct water is not fully recovered. Drought may therefore damage plants irreparably, leading to yield loss, even when they survive the drought itself.

In natural ecosystems when one plant species encounters hard times another may flourish. The most successful animals within those ecosystems will therefore be those which can adapt their food sources according to availability. In traditional agricultural systems human ingenuity could emulate this, so that there would always be something to eat, provided only that it rained sufficiently at some period of the growing season. In modern monocultural systems, however, this advantage has clearly been lost.[42]

Further environmental consequences of drought can include the erosion or destruction of wildlife habitats and a shortage of food and of drinking water for wild animals, with a resulting increase in the incidence of disease. Where possible, animals may react to these stresses in the same way as humans: by migration. Being generally less adaptable in terms of diet, as well as the aptitudes required to find food and water when sources of these change, they may be inclined to migrate more readily than big-brained, resourceful, omnivorous bipeds. On the other hand, they may also be in less of a position to do so, not least because of the way those same bipeds have spread to cover much of the planet's surface. Even if other suitable habitats are in theory available, land-bound creatures may find that corridors to them are not. Drought will therefore, almost by definition, lead to reduced biodiversity, not only locally, in the areas affected, but globally as stresses increase on already endangered species.

Animals also vary in the extent to which they can survive lengthy periods without water. Human beings dehydrate rapidly, and if we lack a mere 11 per cent of our usual water content we are completely immobilised. In general our domesticated animals share this weakness, though there are a number of celebrated exceptions in the camel and its relations. We are, however, at least able to adapt to a very wide range of environments, helped, like rats, dogs and pigs, by our omnivorous appetites. Few wild animals have developed such adaptability. To take an extreme example, not only the animals which live fully or mainly aquatic lives, but also wading birds which rely on bodies of water to breed and feed their young at certain times of year, may be wiped out by a change in rainfall patterns, by land drainage, or by unusual heat. Less obviously, in Britain drought is responsible for declining numbers of birds which depend on the moist soils and small bodies of water that encourage the proliferation of the insects on which they feed their young. Species cited in a recent Royal Society for the Protection of Birds (RSPB) paper on drought in England and Wales include the song thrush and yellow wagtail, but the process is paralleled in many other countries with similar tragic results for species which fill the niche occupied by those birds in England.

To the direct consequences of thirst and loss of food sources must therefore be added loss of habitat. The RSPB, which interprets its remit widely in view of the interdependence of different plants and animals within ecosystems, cites a number of other species which are threatened by drought. They include three species of bat which depend on watery places for their preferred foods, as well as water voles and insects such as mayflies, damsel flies and dragonflies. Amphibians, suffering global population declines which in some cases seem unrelated to drought, are placed under further pressure in England and Wales by declining rainfall. The RSPB is encouraging its nation of gardeners to build and maintain ponds to compensate for habitat loss, but such simple and potentially effective solutions are clearly unavailable in parts of the world where people have less leisure and fewer resources, or when the wildlife in question is not the kind you would want to attract to your back garden.[43]

Loss of a single species would be tragic in itself, but the tragedy goes much further than sentimental or even practical reasons. It is now widely recognised that ecosystems are hugely complex webs of interdependencies. Remove a single element and the ecosystem will change in ways which are often unforeseen. The loss of temporary bodies of water can lead to a reduction in or elimination of the insect species which depended on those bodies of water appearing at certain, predictable times of year. This in turn will lead to a reduction of food sources for amphibians and reptiles, and through this, as well as directly, for birds and mammals. Each of these links in the chain will have created a complex link of dependencies, the disruption of which can again have a range of unpredictable effects. We most often hear of this in relation to the deliberate or accidental introduction of alien species into environments, rats and cats on islands which previously had no land-dwelling carnivores, for example, or the arrival of the rabbit or cane toad in Australia. But the disappearance, decline or upsurge in numbers of a certain species as a result of drought can have similar consequences.

The Socio-economic Consequences of Drought

The historian Michael McKernan began his professional life researching the social impacts of war on the Australian people. When, as he was reading an old newspaper as part of his research, he came upon a farmer and veteran who claimed that drought was far worse than anything he had seen in the Great War, he was at first astonished. As he read further, however, it came to him that drought

> was not just about business, family or personal failure. It was like war. It was about despair, depression, powerlessness, a sense of the unfairness of it all, of life so unpredictable and cruel that it would force fathers to suicide, mothers to depression, children to anxiety.[44]

Perhaps in some ways drought is worse, for in war there is usually a visible enemy whose motives might be explained. Drought, an impersonal force, must seem like a malevolent intelligence, and not just to people whose cultures incline them to such superstitions. It touches every area of life and is capable of unlimited destruction.

In the most systematic studies of the problem yet conducted, the social consequences of drought have been found by Australian researchers to be both destructive and extremely wide-ranging. Rural areas studied by a number of separate teams of researchers during the first few years of this century were found to include income erosion for farms and small businesses, with consequent poverty and the need to seek alternative sources of income; increased workloads both on farms and elsewhere, including on providers of vital social and medical services; an increase in physical and mental health problems and declining educational access. Unsurprisingly levels of out-migration rose, though the areas studied, and rural Australia in general, have long suffered from population loss.

Principal amongst these studies was a report prepared by Margaret Alston and Jenny Kent and published in 2004 by the Centre for Rural Social Research at Charles Sturt University in Wagga Wagga, New South Wales. *Social Impacts of Drought: A Report to NSW Agriculture* noted that nationally, potential economic losses were estimated by the federal government at A$2 billion, much of it directly from major crop and stock losses. The government believed that 40,000 jobs had been lost as a direct result of the drought, while the massive drop in farm income and its impact on small businesses serving rural communities threatened to bring about a wave of bankruptcies.[45] In some cases, well-meaning measures such as the distribution of food hand-outs to needy farming families backfired, driving already hard-pressed small, privately-owned rural supermarkets further towards bankruptcy. Even where bankruptcy was avoided, staff had to be laid off. Whilst this may have troubled small shop owners on a personal level, its impact was potentially greater for small employers generally more in need of skilled or trained workers. Layoffs would mean that such workers were likely to move to other areas of the country, creating fear of a skill shortage when hard times abated.[46]

The loss of people from communities can have particularly demoralising effects, as two residents of one small and badly-affected rural Australian town told researchers. The town's mayor noted that the loss of residents meant that 'all the numbers start deteriorating and the government agencies look at their support as far as their staffing levels go too and there's a bit of a snowballing effect'. Another resident, identified as a 'Business Person', described the knock-on effects of drought in a way which recalls the interdependencies of natural ecosystems:

> Schools don't need as many teachers because there's not as many kids there, so more people leave town. Because they are not shopping in the stores one of the stores will close down. Once that starts it's very hard to arrest.[47]

Alston and Kent found that, in the particular context of rural Australia at the beginning of the twenty-first century, it was difficult to sort out the effects of drought from those of an ongoing economic transformation – or rather, decline – which they call 'structural adjustment'. They felt, however, that the drought had at the very least accelerated processes already evident. In the conclusion of their report, which was commissioned by the New South Wales government, they identified a number of clear impacts, problems which had either been caused or exacerbated by prolonged drought. Loss of population had accelerated, especially in relation to young people. Economic strain was felt on farms, particularly in the areas dependent on irrigation, and in the businesses to a greater or lesser extent dependent on them. This had resulted in a reduction in workforce sizes, while on the other hand many individuals suffered increased workloads. All of these factors fed into a reduction in voluntary activities and community participation.

Alston and Kent talk about a decline in 'social capital', which is defined as 'networks of trust, reciprocity, social norms and proactivity that create the glue that holds a community together'. Informal and formal community networks 'become more fragile'. Landcare, a state-sponsored volunteer-based organisation which organises environmental and farmland improvement work through local groups, was in decline, along with similar volunteering activities, with organising committees meeting less frequently. Charitable organisations were finding it hard to attract volunteers, just when many of them needed them most. This decline is part of a broader picture in which 'individual and community resilience has been sorely tested by the drought and also by the policy parameters that shape drought response'. Alston and Kent give the example of small businesses having difficulty accessing assistance, and farming families facing overly complex paperwork in order to get benefits to which their situation entitled them. As a consequence, as communities and individuals threatened to buckle under the pressure, professionals and auxiliary workers whose job it was to help these communities and individuals felt the strain themselves. A grim feedback loop was also in play here, as many workers in such places as hospitals and schools were themselves part of farming families who were in the drought's immediate line of fire. Not only did this mean a double burden of strain, but the fact that they were still earning might serve to disqualify their families from benefits. Alienation and mistrust of government, characteristics of Australian rural life during its long economic, social and demographic decline, have been exacerbated. The overall picture is of communities, families and individuals proud of their self-reliance soured by being placed in a situation where self-reliance is simply not enough. Feeling that they had made an enormous contribution to creating a prosperous and successful Australia, they now felt humiliated by Australia's response to their plight. Even the generally optimistic 'Business Person' quoted in the report's conclusion admitted that people 'sometimes might sort of be a bit harder to get on with'.[48] Alston and Kent wrote their report in 2003, but the drought was to persist. In 2007, farmers held the highest rate of suicide of any group of Australian, 96 per cent of New South Wales was in drought and severe water use restrictions were imposed through most of the country.[49]

The impact on individuals varies as well, of course, depending on their level of existence before the drought and factors such as age, gender and support network. These problems, tellingly, have been compounded by what one of the groups of researchers identifies as 'structural adjustment, the loss of services and an economic rationalist focus on service and infrastructure support'. Such impacts spread, particularly through increased migration, to areas where the direct effects of drought are less, particularly cities.[50]

Clearly, the economic consequences of drought overlap with these social consequences, to the extent that the two are intertwined and often difficult to separate, were there any point to such an exercise. Industries and modes of agricultural production can be undermined to the point of destruction, causing unemployment and, at best, eventual economic transformation. At worst, the removal of a society's economic base may prove irreversible and the base itself irreplaceable.

The Intergovernmental Panel on Climate Change notes the following likely economic effects of drought which it classifies under four headings: under 'agriculture, forestry and ecosystems' are listed land degradation, lower yields, crop damage, crop failure, increased rate of death among livestock, increased risk of wildfire; under 'water resources', simply 'more widespread water stress'; under 'human health', an increased risk of food and water shortage and of malnutrition and water-borne and food-borne diseases. Under the final category, 'industry, settlements and society', it lists water shortages for settlements, industry and societies, reduced potential for hydropower generation and an enhanced potential for large-scale migration.

As a developed country Australia has immense resources at its disposal, not least of which is the level of education and the general health of its population. The difficulties will be compounded in other parts of the world by poverty and underdevelopment.

In the Somali region of Ethiopia, for example, the death of farm animals due to prolonged drought has reduced the incomes and direct access to food sources of families already surviving on scarcely adequate resources. The need to compensate for lost revenue means that children, in a region where only one in five was enrolled in school in the first place, are forced to leave school, substituting short-term survival for a longer-term route out of poverty for themselves and their families. Some find work as domestic servants, while others are forced to work informally, for example as shoe cleaners, while some can only scavenge for food. By May 2006, reduced rolls had forced eight of the 31 schools to close. Children able to continue attending school may be affected by hunger, damaging their ability to learn. 'The education system in Ethiopia's Somali Region was under serious stress, even before the drought,' according to UNICEF official Augustine Agu.[51] It had some of the country's lowest enrolment rates – 21 per cent overall against the national average of 79 per cent.

Many of the region's people are herders, which means that one response available to them is to drive their livestock to other regions, where they hope to find water, though the potential for social conflict in such a scenario is clear. According to Augustine Agu, the 'insecurity and collapse of social infrastructure' provoked by the drought 'forced entire communities to abandon their homes'.[52]

In Australia, however inadequate government response may be, or be seen to be, there is at least a functioning authority with recognised responsibility for addressing the situation. In Ethiopia and many other developing countries, such a body cannot be said to exist. The efforts of UNICEF and associated NGOs in the Horn of Africa are clearly aimed at compensating for this absence, but given the constantly recurring crises, many of them drought-associated, their attempts to move beyond palliative measures to address long-term infrastructural inadequacies are constantly frustrated. Even when the rains eventually came to Ethiopia, chronic problems of poverty, disease and malnutrition remained. In Australia, too, rains will not solve the problems of long-term rural decline. Communities there are being destroyed more slowly, and individuals can survive by moving to cities which offer employment and other opportunities. In Ethiopia, few such alternatives exist. In both cases, drought served both to exacerbate and to draw attention to long-term structural problems which rain, when it arrives, does not wash away.[53] As the OECD observes, 'One should differentiate between countries experiencing *physical* water scarcity (e.g North Africa and the Middle East) and those affected by *economic* water scarcity (sub-Saharan Africa)' (emphases in original).[54] If a country or region lacks water, but enjoys a surplus of economic resources, it can import food. If it is experiencing 'economic water scarcity', however, shortage of water *per se* is not the problem. A lack of infrastructure and general economic resources enabling the country's people to make optimum use of the water creates scarcity, or the impression of it, which if you are a thirsty child or a thirsty cow amount to the same thing.

In North Africa, too, though governmental structures are generally intact, and far more developed, the amount of water available per head is falling and social conflicts arising accordingly. In such situations, a failure of already inadequate levels of rainfall can have the same devastating effects as we have seen in Australia and Ethiopia. In these countries, which are neither 'developed' nor 'least developed', the impact of poor or deliberately exploitative policy decisions is often most evident. Australia, after all, is rich enough to cope, and so policy decisions can be expected to be about the best means to do so. Ethiopia is so poor that all that seems immediately available is some kind of palliative action. In Algeria, however, it is quite clear that a strong, functioning, participatory state could go a long way to improving the country's resilience in the event of drought, not to mention its water supply under normal, admittedly generally arid, conditions. Here the problems seem less to do with nature, even with climate change, than they are rooted in bureaucracy, incompetence, a failure to keep pace with demographic growth and, increasingly, the prioritisation of those sectors of the economy favoured by foreign capital and IFIs.

In other words, underlying the water supply problems of Algeria and its neighbours is that same disastrous synergy: the spread of neo-liberalism at a time of increasingly apparent climate change, widening disparities of wealth and income and growing populations. Rostomi Hadj Nacer, a former governor of Algeria's national bank, attributes his country's water supply problems to the 'lack of economic administration, the shortage of experts' and to 'incompetence and short-term management of government stock' which mean that 'asset-destruction is reproducing itself ad infinitum'. The reason, in other words, for the exacerbation of the problem is that just when Algeria 'needed the state to become stronger' the opposite happened. In this man's view, economic liberalism, rather than necessitating a weak state, means on the contrary that you need to strengthen it.[55] This is not a view which the authors share, but it has its own logic: if you are going to let the gangsters loose, at least make sure the serious crime squad is fully staffed and well-trained.

Planning for Drought

Drought management 'involves decision making under uncertain conditions. It is risk management.'[56] In many parts of the world it requires contingency plans for an event which may never occur, and those responsible complain that it can be difficult to get the attention of people and their political representatives until it is too late. Drought planning invariably involves coordination among multiple agencies and each of these agencies is likely to have other, more urgent, priorities. Daniel Loucks, an American expert on drought planning, complains that drought response tends to be crisis-oriented, with structures being hastily constructed or adapted in the face of emergency. As a consequence they are 'largely ineffective, poorly coordinated, untimely, and inefficient in terms of the resources allocated'. He proposes using advanced computer technology in order to draw up and implement detailed plans which should:

- specify a sequence of supply- or demand-side measures which will increase in stringency as drought intensifies;
- lay down guidelines designed to minimise the impact of such measures on health, economic activity and the environment;
- include provisions for forecasting drought conditions;
- include drought indicators and drought triggers;
- take into account the political and social environments in which they will be implemented;
- involve and educate the public;
- accompany penalties for non-compliance with economic incentives to conserve water.[57]

This thinking does seem to be finding a response from policy-makers in various parts of the world. When researchers established that the Colorado River was suffering from the effects of over-allocation, the Southern Nevada Water Authority began to offer residents of the area financial incentives to reduce their water use, such as paying them to remove their lawns or parts of their lawns, use car wash services which waste less water and install covers on their swimming pools to reduce evaporation.[58] The campaign was backed by a colourful campaign of humorous TV ads under the slogan 'Don't make us ask you again. It's a desert out there.' On 4 June 2008 Arnold Schwarzenegger, Governor of neighbouring California, declared the state to be officially in drought and threatened draconian water restrictions if people did not respect proposed voluntary conservation measures. The drought package included emergency measures such as shipping water into particularly stressed areas, but Schwarzenegger's long-term plan is to use technical assistance to make irrigation systems more efficient, combined with restrictions on domestic water use, in order to cut consumption by 20 per cent by 2020. His plan seeks to avoid upsetting powerful agricultural interests by imposing most demands on domestic users. Denying planning permission to housing developments which cannot guarantee a 20-year water supply is precisely the kind of measure which is needed, but the inability to tackle agriculture with a stick to go with the carrot of state aid to improve irrigation efficiency will be the plan's downfall. Once again, the profit motive will have triumphed over urgent necessity.[59]

Though it lacks the legal competence to enforce them, the European Union, through the European Commission, would seem to be ideally placed to aid and encourage its member states in drawing up similar plans, though they would be just as likely to be stymied by corporate agribusiness as they have been in California. In Chapter 5 we will see how well the EU fulfils such tasks.

Desertification

If drought continues for long enough, and measures such as those outlined above prove ineffective, it ceases to be drought as such. Drought is by definition an event and drought as an event is by definition an exception to the normal level of water availability. A permanent drying of the weather is simply a change in the climate and it can lead to the development of desert conditions where previously there was an adequate and predictable supply. The definition offered by Constance Hunt is:

> the degradation or destruction of land to desert-like conditions that can include the growth of sand dunes, deterioration of rangelands, degradation of rainfed croplands, waterlogging and salinization of irrigated lands, deforestation of woody vegetation, and declining availability and/or quality of freshwater.[60]

When it rains, the water which falls on vegetated areas – in other words, almost anywhere which is not either already desert, frozen waste or 'built environment' – is either taken up immediately by plants, or is absorbed by the soil, or evaporates or runs off. The precipitation which is taken up by plants is returned to the atmosphere in the evotranspiration process, whereby plants draw water from the soil, which then evaporates. That which is not may be stored in the soil and used later by plants or other soil-dwelling organisms or it may seep into underground aquifers. Photosynthesis is enabled by this stored water and the energy provided by sunlight, but in conditions of aridity this process is changed by the fact that one of its essential foundations – water – is in short supply while the other – energy from sunlight – is abundant. Vegetation becomes sparser, and even during the infrequent spells of intense rainfall characteristic of many arid regions, water tends neither to soak into the soil – unless it has been hoed to break up the relatively impervious crust formed during long periods of intense sunlight, or, to put it another way, subject to human ingenuity – nor to run off into rivers, lakes or aquifers, but to be lost to evaporation.[61]

Drylands can be self-sustaining ecosystems just as well as wetlands, but human activities can in both cases upset the natural balance. Land which is cleared of vegetation and not planted and hoed will form a crust, preventing seepage. The soil below will therefore dry out, while aquifers beneath may be depleted. Runoff accelerated by the lack of absorbency of the top soil horizon will generate erosion. The result is a downward spiral into desertification, in which soil nutrients are either washed away or blown away. Where agriculture is rain-fed rather than dependent on irrigation, land left fallow is vulnerable to these processes of water and wind erosion. On irrigated land, inadequate leaching of salts either naturally present in the soil or contained in water used for irrigation leads to salinisation, often accompanied by waterlogging.[62]

Desertification is rarely if ever the result of purely natural processes, or even of climate change. Often a snowballing process, it is a result once again of synergies: razing of vegetation, the resulting destruction of other plant and animal life and accelerating erosion and removal of sediment which in turn is deposited on fields. Although areas are most vulnerable if they are agriculturally marginal to begin with, and although climate change is clearly enhancing the vulnerability to it of numerous areas, desertification is a product above all of over-exploitation. Desertification is a growing problem in rural Australia, the African savannah, the Asian steppe, the North American Great Plains and South American Pampas, often places where existing desert meets grassland.[63] Less well-known is the threat it poses in much of the Mediterranean region of Europe. Here the problem is quite starkly one of mismanagement rather than anything to do with declining availability of water. A decade ago the EU's official soil monitoring body reported that erosion was at the root of spreading desertification in the European Mediterranean and that 'already large areas have been so severely damaged that they are no longer capable of supporting arable agriculture. This is leading to depopulation of the land.' While erosion due to what the report called 'unsustainable agricultural practices' was a continent-wide phenomenon, in the hotter, dryer countries of the south its impact was more severe. 'Much of the Mediterranean zone of Europe is now threatened by desertification', the report concluded.[64]

How well the European Union and its member states are responding to this crisis will be considered in Chapters 5 and 6.

2

FLOOD

Once a rosy lawn,
But now a muddy pond.
Floods in fields of cotton and the sugar cane...
I'm so weary, heavily down and blue,
With no one to tell my troubles to.
I'm just a wondering, homeless,
lonesome refugee.

Lonesome Refugee, Laura Smith[1]

Floods clearly present very different problems from droughts, which is as much because of their generally sudden impact as it is to do with their status as opposites. As the Jucar Basin researchers cited earlier note, 'Drought's importance lies in its slow and progressive nature, which makes Basin managers deny the event until they are completely inside it.'[2] This would be difficult to do in the case of flood, though there are certainly decision-makers one could imagine crying 'it's nothing but a heavy shower' as they float away on their desks, surrounded by their office pot plants, never to be heard from again. There are planning authorities, for example, like the one in England which gave permission to build houses on land certain to suffer flooding within the lifespan of the buildings and then failed to maintain flood defences.[3] In the Netherlands, on the other hand, avoiding building in flood-prone areas is scarcely practical. Just over a quarter of the country is actually below sea level, but seven-tenths of it would be flooded without some form of coastal defence. Moreover, the quarter which is below sea level is responsible for a massive 70 per cent of the Netherlands' GDP.[4] Short of evacuating the country, the

Dutch must find solutions to the constant threat of flood which they face. Given the Dutch people's long experience with flood and flood protection, the rest of the European Union has much to learn from them, especially at a time when sea levels are expected to rise. We will look at some of the solutions which the Netherlands has found to these problems later in the chapter.

In the last decade of the twentieth century the UN estimates that 665,000 people globally died in over 2500 natural disasters, 90 per cent of which were water-related. All bar a few thousand of the victims were from developing countries. Flooding was responsible for only 15 per cent of deaths due to natural disasters, and drought 42 per cent, though half of the natural disasters involved flooding.[5] This was partly because of the recent growth of settlements of all sizes close to the sea. Climate change also played its role, as 'increases in heavy precipitation events' occurred, and the view of the Intergovernmental Panel on Climate Change is that, although natural variations in rainfall patterns make quantification extremely difficult, there is no doubt that these events 'are associated with increased atmospheric water vapour... and are consistent with observed warming'. This increase has also confirmed projections based on theoretical and climate model studies which 'suggest that in a climate that is warming due to increased greenhouse gases, a greater increase is expected in extreme precipitation, as compared to the mean'. The IPCC also predicted that this trend would continue, with both frequency and intensity of 'heavy precipitation events' set to go on increasing in tropical and subtropical zones as general levels of precipitation rise.[6]

As well as living and gaining their livelihoods in areas subject to flood, people have increased their vulnerability by effecting changes to watersheds, to rivers and coastlines, to agricultural practices and to landscapes. Deforestation or the removal of other natural, perennial vegetation, and even poorly-designed systems to combat flooding, can all contribute. Ill-thought out or wholly unplanned urban developments can be vulnerable not only by being built in the wrong place, but by increasing the proportion of water which will find its way directly into watercourses during rain. The destruction of wetlands removes natural barriers to flooding, as does deforestation.[7] The OECD sees this largely as a consequence of policy failure, noting that:

> While intense rainfall is a natural occurrence, the magnitude and velocity of the ensuing large water flows are affected by human actions, as is the vulnerability of human settlements to flooding and erosion. Natural flooding has been exacerbated in many cases by fragmented responsibilities and lack of integration of policies relating to flood protection, land use planning, and flood damage compensation. Even where coherent policies are in place, land use and building height restrictions in flood plains are not always respected, and compensation payments may even permit property owners to return to the pre-flood situation that led to the damage in the first place.[8]

But before examining the socio-economic effects more closely, we need to take a brief look at flood's environmental implications.

The Environmental Consequences of Flood

Plants exhibit a wide range of responses to flood, which from a plant's point of view is defined as saturation of the soil and submergence of the plant's roots, as well as deeper inundation in which most or all of the plant ends up under water. The plant's situation is in one way, at least, comparable to our own. While drought is for the most part a condition to be avoided and, if that should prove impossible, survived, flood may be beneficial. There are even species which depend on it to thrive. Many others are perfectly able to survive flood, either as individuals or by arranging, by one mechanism or another, to pass their genes on to the next generation. Given the range of habitats which plants have adapted to, and the range of niches within those habitats which they are able to fill, it is unsurprising that this should be the case. A special edition of *Annals of Botany* devoted entirely to this subject set the scene in a way which succinctly captured the range of possibilities. 'Flooding stress is... a strong driver of adaptive evolution' said the authors, and:

This has resulted in a wide range of biochemical, molecular and morphological adaptations that sanction growth and reproductive success under episodic or permanently flooded conditions that are highly damaging to the majority of plant species. However, even seemingly poorly adapted species possess some short-term resilience that is important for overall success of these plants in various habitats. Stress on plants imposed by flooding of the soil and deeper submergence constitutes one of the major abiotic constraints on growth, species' distribution and agricultural productivity.[9]

Animals, even more obviously, will respond to flood in a wide variety of ways, depending on the species. Ecosystems may thus depend entirely on periodic flooding, as in Europe's ancient water meadows or large parts of the Amazon forest, or they may be entirely destroyed by flood as an unprecedented event. Ecosystems may also be vital in preventing flooding, so that upsetting their balance may actually precipitate flood events.

Three examples come to mind. Firstly, the Grand Canyon on the Colorado River in Arizona: on three separate occasions since 1996, the Grand Canyon has been deliberately flooded in an attempt to restore at least in part the ecosystem which has been undermined since the building of the Glen Canyon Dam higher up the river.[10] Secondly, New Orleans, where human alterations to the course of the river and the removal of coastal vegetation turned what would in the past have been at worst a series of localised tragedies into a catastrophe of international significance.[11] Thirdly, the removal by land-use change of floodplains along Europe's great rivers, such as the Rhine and the Danube, which has increased the continent's vulnerability to large-scale flood damage.[12]

In each case, human activities affected vulnerability to flood. Vulnerability can increase either because flood protection measures turn out, in the long term, to have been ill-thought out, as in New Orleans; or because infrastructure built for other purposes – primarily power generation, water storage and navigation – has not taken increased flood risk into account. This might be because at the time the works were undertaken the full extent of the problem was not understood; or it might be that a country or region prioritises its own needs, acting with indifference to the interests of downriver neighbours; or it might be because short-term profits trump, as so often, long-term interests, or powerful private interests take precedence over those of the broader public. Whatever the reason – and often a mix of more than one of these factors is at play – the results can be catastrophic, not only for the environment but for the lives and livelihoods of the people affected.

The Socio-economic Consequences of Flood

Floods, if they are seasonally predictable, can become the basis of agricultural systems, a phase in the life of hunter-gatherer or herder communities, or a source of water which can be stored for the dry months which will follow. Only when they become unpredictable or come into conflict with human activities do they start to constitute a problem. As Constance Hunt writes, 'A flood hazard is the threat to life, property and other valued resources presented by a body of water when it rises and flows over land that is not normally submerged. A flood is not hazardous unless people are somehow adversely affected by it.'[13]

The IPCC identified a number of likely economic effects of flood which will become evident as heavy precipitation events increase in most parts of the world, a scenario it classifies as 'very likely'. Using the same broad classification as it did for drought, the IPCC asserts that 'agriculture, forestry and ecosystems' will suffer from damage to crops, soil erosion and waterlogging. Water resources will be affected by deterioration in quality and contamination of the water supply. Injuries and infectious respiratory and skin diseases will become more widespread and the death rate will increase. The IPCC notes that 'populations with poor infra-structure and high burdens of infectious disease often experience increased rates of diarrhoeal diseases after flood events'. In both developed and developing countries, however, mental health is affected, 'with flooded persons experiencing long-term anxiety and depression'.[14] Settlements, commerce and transport will be disrupted and pressures on infrastructure will increase. Public and private property will be destroyed.[15]

Assessing the economic and social consequences of flood is less straightforward than it might appear. For one thing it is not sufficient to measure the value of damaged property in order to assess the economic harm done by a flood. Leaving aside the contentious matter of the 'economic value' of a human life, a distinction must be made between the loss of simple goods and the loss of productive capacity. To take a straightforward example, if a teddy bear is destroyed it may result in tears from its owner, but it will not have any impact on GDP or on the productive capacity of a household. If the sewing machine on which it was made by the owner's grandmother is destroyed it will affect the latter. If the factory in which the sewing machine was manufactured is washed away it will affect the former. So economic value should not be measured merely by the cost of initial purchase, or the replacement cost, but by its capacity to contribute to production. Both kinds of loss are important, but in the aftermath of a destructive event it is the loss of productive capacity which is the more significant.

One interesting study of the social and economic problems resulting from flood was conducted in 2008 by the University of Iowa, following damaging floods in this largely agricultural region of the United States. Researchers constructed a rough taxonomy of losses and damage due to the floods, beginning with households, where losses 'include personal items, household goods, vehicles, homes, and in some cases, lost wages or even lost jobs'. Farmland is damaged in the form of flooded fields and lost crops. In Iowa, this was reduced because of the state's two principal crops – corn (maize) and soya beans – the latter could be replanted, the floods occurring early enough in the growing season to make this possible. In both cases – households and farms – insurance limited individual losses to a certain extent. Insurance and direct compensation, however, while they spread loss, do not necessarily actually reduce it. Whether they do so or not depends on a large number of variables, but certainly some loss will remain. Businesses, too, will suffer lost stock, lost sales and lost profits. The Iowa researchers predicted that as the national economy was in any case slowing, there would be numerous bankruptcies and job losses which could be directly attributed to the flood.

Losses were also felt by communities: public service delivery was adversely affected and in come communities 'waste water and fresh water facilities are compromised and must be restored'. In addition, there was a need 'to repair roads and bridges, public lighting, public parks, and public buildings.... (and) to create additional solid and hazardous waste disposal facilities'. Again, some of the funds required for this might be accessed from outside sources, in this case 'federal and state disaster assistance', but while this is clearly of importance to the communities immediately affected, it does not change, at least not directly, the overall effect when measured against national GDP. Finally, the researchers did not neglect to mention 'intangible assets', although the economic value of these is by definition impossible to assess. They mention, for example, 'the breakup of established neighbourhoods and the loss of iconic stores or other institutions that provided local color and character'.[16]

As with the example of the Australian drought, residents of the part of Iowa affected by the floods at least have the advantage of being members of a prosperous, developed society with recourse to functioning governmental structures and a range of resources. When floods hit much poorer countries the effects can be much more severe and recovery much slower or even impossible to effect. Bangladesh is, for example, one of the world's most flood-prone nations, a low-lying country occupying the flood plains of four major rivers. In 1999, following even more severe floods the previous year, tens of thousands of hectares of the country, including productive farmland, were completely inundated, and hundreds of thousands of people were forced to abandon their homes. On one level, damage can be taxonomised in much the same way as that occurring in Iowa, but the scale of it was much worse and the loss of life far greater. Typically, flood-related deaths in a developed country tend to be from drowning, collision or exposure, whereas in poorer countries added to these causes is an upsurge in water-borne diseases such as diarrhoea. If a hospital is affected by flood in a developed country, alternatives usually exist, whereas in a country such as Bangladesh this is unlikely to be the case. The same goes for other public facilities such as schools, so that education will suffer much greater disruption, the effects of which may be felt for decades.

Whereas the response of the authorities to flood in a developed country may be inadequate or flawed, in a poorer country there may be no effective response whatsoever. Of course, there is a glaring exception to the first part of this statement: New Orleans. New Orleans, however, was precisely that: an exception. It is impossible, in the history of floods in developed countries to find anything approaching the astonishing incompetence, indifference or mendacity of the response of the US federal authorities to the destruction of one of the nation's great cities. When neo-liberalism is dead and buried and we are dancing on its grave, the New Orleans flood will serve to remind us that it was a philosophy whose fundamental tenet is that the poor, when not being exploited for gain, should be left to their fate.

One of the worst floods that Europe has seen in living memory was the one which hit the Netherlands in 1953, covering 200,000 hectares of land, killing 1836 people and 200,000 farm animals. When sea dikes were breached, 3000 houses and 300 farms were destroyed and another 40,000 houses and 3000 farms damaged. 72,000 people had to leave their homes and were evacuated to other areas. Salt water contamination made huge areas of farmland infertile for many years afterwards.[17]

The storm, which was precipitated by gales of up to force 11 which caused huge waves across the North Sea, also created extensive damage and some loss of life in the United Kingdom and Belgium. Any repeat of this event would undoubtedly cost more lives and bring about far more costly damage. In 1953 the Netherlands, though no Bangladesh, was a relatively poor country recovering from war and occupation. In the early twenty-first century it is one of the world's most prosperous countries, densely populated by over 16 million inhabitants. It is, moreover, one of the power-houses of the European economy, so that the consequences of any repeat of 1953 would be felt across the continent and beyond. Investment in flood protection is therefore one of the country's least controversial priorities. The question is, what form of flood protection should the Netherlands and other vulnerable regions of the world be investing in?

Flood Control – Good, Bad and Indifferent

The principles of flood control can be understood by studying the single catastrophic case mentioned above, that of New Orleans. It is misleading to say that the New Orleans flood was 'caused' by Hurricane Katrina. The real causes of the flood were a familiar panoply of corporate greed and human failings: hunger for profit, misunderstandings, mistakes, folly, hubris and corruption, with in this case a strong dash of racist and class-based indifference to the fate of the dark-skinned and the poor. Houses and infrastructures were built where they should not have been built; the nature of the Mississippi River as a group of ecosystems was ignored and some of the structures intended to contribute to flood control actually had the opposite effect.

For example, as an ecosystem which provides natural defences against flood, the Mississippi Delta needs a constant supply of sediment. Much of this sediment originates in the Mississippi's longest tributary, the Missouri. In the 1950s several big reservoirs were built on the upper Missouri River, the aim of which, apart from enhancing the water supply, was to protect downstream areas from floods. Unfortunately, they also trapped sediments, so that, as Richard E. Sparks, an expert on the ecology of large floodplain rivers, explains:

> Much less fine sediment (silt and clay) flows downstream to build up the Delta during seasonal floods, and much of this sediment is confined between human-made levees all the way to the Gulf, where it spills into deep water. Coarser sediment (sand) trapped in upstream reservoirs or dropped into deep water likewise cannot carry out its usual ecological role of contributing to the maintenance of the islands and beaches along the Gulf, and beaches can gradually erode away because the supply of sand no longer equals the loss to along-shore currents and to deeper water.

The result was that Hurricane Katrina was far more destructive than it would have been had it followed the same path 50 years earlier 'because more barrier islands and coastal marshes were available then to buffer the city'. The beaches and islands allowed vegetation to develop and vegetation acts as a natural barrier, reducing the height of waves and holding the shoreline relatively intact. It is also self-regenerating and therefore 'in contrast to a cement wall... would recolonise and repair a breach in the dune'. Best of all are trees, but these grow best at higher elevations, as unlike marsh grasses they will not tolerate frequent flooding. 'The greatest resistance is offered by tall trees intergrown with shrubs; followed by supple seedlings or grasses; and finally mud, sand, gravel, or rock with no vegetation.' Unfortunately, navigation canals cut into the Delta disrupted the vegetation, causing salt water to flood the coastal freshwater marshes.

The first step in this particular flood control exercise should therefore be to attempt to restore the vegetation and study how the natural system can be reconstructed. But there are further problems. The Upper Mississippi Basin houses much of the US corn belt and up to 90 per cent of the original wetlands of that region have been drained. The spread of the impervious surfaces which characterise cities, towns and suburbs; agricultural systems designed for rapid drainage to allow early planting; and the construction of levees along all of the Mississippi's tributaries' floodplains, all increase the height of floods downstream.[18]

Sparks suggests numerous ways to alleviate these problems and thus prevent the occurrence of a second New Orleans, narrowly avoided when a further hurricane veered away from the city at the last minute in August 2008.[19] These include restoration of some of the wetlands in the watersheds; changes to building materials so that they soak up more water and the incorporation of plants into rooftops; reconnecting some floodplains with their rivers and learning to live with the results. In and around New Orleans itself low population density means that large areas of land on the Delta could be used as a buffer. The enormous expense of building the kind of flood protection which keeps the Netherlands from flooding could therefore be avoided. Sparks does, however, advocate a 'Dutch approach' to other aspects of the problem: he praises the 50-year time horizon which characterised the planning effort with which the country responded to its own catastrophic flood in 1953, and the commitment where possible to working with nature, rather than against it.[20] Unfortunately, all economic activity in the United States is conditioned by the methodology characteristic of corporations, and corporations do not deal in half-century timescales.

A limitation of any plan for flood control is the ability to predict the height of future floods. The likelihood of a flood occurring is expressed in 'years'. A '100-year flood' is one which is likely to occur once per century. Thus the chance of it occurring in any given year is 1 per cent. A 200-year flood has a 0.5 per cent chance of occurring in any given year, and so on. Unfortunately, the estimates of flood probability are often deeply flawed. Firstly, because very few areas have reliable records going back a hundred years or more; and secondly, because changes in landscape, land use and the river itself make past records a poor guide to the future. Many of the changes that have occurred in Europe and elsewhere in modern times have increased the likelihood of flood. As we have seen from the example of New Orleans, these may even include well-intentioned attempts to reduce this risk, for we are dealing with complex natural systems, changes which can produce results utterly different from those sought or predicted. Many, however, are simply the result of population growth, prosperity and 'modernisation'. These factors played a role in New Orleans: the transformation of wetlands, forests and other natural landscapes into farmland; the spread of human habitation and infrastructure; projects to enhance the navigability of rivers; and the replacement of systems of agriculture which incorporated and profited from flood with those which rely on rapid drainage. On top of all this, the building of levees and other structures to control floods hugely enhances the dangers in the event of a flood so high that the structures fail. As Hunt explains:

> under unaltered conditions, floodplain land is subject to gradual inundation as the flood event reaches a peak and then recedes. In contrast, flood-

control structures are most likely to fail during the flood peak. The sudden torrent of water that is released when a dam or dyke fails can be much more destructive than a more natural flood event.[21]

Trying to control a flood by building barriers to hold back water is in most cases ultimately a doomed enterprise. Such structures interfere with natural patterns of drainage. They can actually cause or exacerbate flooding by stopping the water returning to the water course after a flood recedes, or by preventing natural drainage of water from precipitation. Pumps or sluice gates can be a solution but these are expensive. Sophisticated flood prevention systems such as those in place in London or Venice can only be afforded in the most exceptional circumstances. Part of the point of barriers is to divide rivers from their floodplains, but the natural storage represented by those floodplains is not replaced. The result, as with New Orleans, is to increase the risk of flooding further downstream. This problem can also be addressed, provided the whole river is managed by a single authority capable of overriding local decisions. Yet, as the example of the Mississippi demonstrates, even within a single country this is rarely the case. Indeed even in the Netherlands local authority planning powers make a national plan difficult to implement. Where rivers pass through more than one sovereign state, the potential for conflict is clear. Even communities facing each other across rivers may take uncoordinated decisions about flood protection, with possibly disastrous consequences. When it comes to places separated by hundreds of miles of river, such decisions may not even be known about until it is too late.[22]

Floods do not deposit only water on adjacent land, but sediment. In many traditional systems of agriculture this is the basis of the land's fertility, but when such systems are abandoned the sediment merely adds to the damage caused by the flood. Yet if it is not allowed to stream out of the river, it does not disappear. Instead, it accumulates on the bottom of the channel. This may cause problems for navigability, but it also increases the likelihood of flooding, or the likely severity of any flood, by raising the elevation of the riverbed, in some cases even above the level of the river's floodplain.[23]

Channelisation is, if anything, even more misguided than the building of levees, which in some cases at least is unavoidable. Channelisation covers a number of types of alteration to a river's natural course. It can include the straightening of that course, the widening or deepening of the channel or the changing of its shape to speed flow, the paving of the channel or the removal of vegetation from a river's banks. All of these processes may be undertaken for a number of purposes, including navigability, but whatever the purpose, one result is likely to be an increase rather than a decrease in flood hazard, as Hunt explains:

> The shortening of the channel in an attempt to speed the delivery of floodwater to some outlet often removes many hectares of floodplain land that would, under natural conditions, provide areas for storing and conveying high flows. Thus, the concentration of the flood volume into a single, straightened channel can result in increased flood peaks.[24]

In addition, the channelisation of tributaries means that natural differences between them will be smoothed out. Water will pass downstream more rapidly and more uniformly and the river into which the various tributaries run will be likely to suffer simultaneous floods along its course, putting pressure on infrastructure and services, including specialised and emergency services. Paving a channel to increase flow rate prevents the natural exchange of water between river and aquifer. If the river discharges into the aquifer, recharge rate will decline, contributing to water shortage and drought. If the opposite is the case the river will lose water volume, which may be problematic throughout its cycle but particularly critical when it is at its lowest.[25]

Removing vegetation, as we have seen in the case of New Orleans, increases flood hazard by removing the most effective barrier, but it also increases the rate at which water runoff enters the river and can therefore increase flood hazard. Yet it is purportedly designed to do just the opposite, again by increasing the rate at which floods will be carried downstream.[26]

Physical Barriers – the Case of the Netherlands

There are, however, circumstances in which physical barriers to flood cannot be avoided. The degree to which wetlands and natural vegetation can protect from flood is clearly limited, as well as being difficult to estimate reliably. As we have seen, without such physical barriers, 70 per cent of the Netherlands would be subject to frequent inundation. Indeed, the Netherlands in the form in which it is known today would not exist. While one of the authors has a clear vested interest in this existence, the other would readily concede that a world without Rembrandt, Vermeer, Van Gogh, Delft pottery, oude jenever, tulips, stroopwafels, Anouk and Johan Cruyff would be immeasurably culturally poorer. How then, can a country which nature intended to be under water keep its feet dry?

The answer is that even in the extreme case of the Netherlands, working wherever possible with nature rather than in a constant battle against it is now seen as the only way to achieve long-term safety. The consensus is that instead of attempting to strengthen engineering structures to cope with rising sea levels, sands should be 'nourished' to encourage the growth of vegetation. At the same time, the flow area of rivers should be enlarged where possible rather than dikes being strengthened. In other words, spaces must be preserved, enhanced or created into which water can flow, in which it can be absorbed and damage to other areas avoided. Care must be taken when new building projects are proposed or approved.[27]

Modern flood defences trace their origins to 1937, when the 'Rijkswaterstaat' (Department of Public Works) published a study which demonstrated the very high level of vulnerability of much of the country to flooding in the event of storms and high sea levels. This led to the 'Delta Plan', under which the mouths of major rivers would be closed off to create freshwater basins, having the beneficial side effect of providing almost limitless supplies of fresh water. The plan's implementation was delayed by war and occupation, but in 1950 the first two dams across the mouths of the Brieles' Gat and the Botlek were closed. Remaining dams were supposed to be completed over the following decades, but the 1953 flood led to a new sense of urgency. Less than three weeks after the flood, in February 1953, the Delta Commission was established with the aim of speeding up the work.

The resulting engineering works are among humanity's most awe-inspiring physical achievements, fully justifying the claim that they are the 'eighth wonder of the world'. A storm surge barrier across the River Hollandse Ijssel, completed in 1958, protected the densely populated western part of the Netherlands. Over the next 30 years, a large number of barriers, sluices and dams were built, creating freshwater lakes behind them and transforming the ecology and economy of the country.

The 'Deltawerken' meant reduced reliance on the traditional dikes, the total length of which was reduced by 700 kilometres. They had a number of other advantages: improving the supply of fresh water, ending the centuries-old isolation (and consequent poverty) of much of the province of Zeeland and enabling the further development of inland water transport.[28]

These tremendous engineering achievements should not be underestimated and are unlikely to be by anyone who has stood in their presence. It might also be noted, *passim*, that it would be ludicrous to imagine that they could have been accomplished by the private sector which, like the rest of the Netherlands, shelters confidently for the most part behind them.

Nevertheless, even in the Netherlands humanity's ability to control nature has its limits and, as noted above, that limit appears to have been reached. Instead of further science fiction masterpieces, the Netherlands' hydraulic engineers now recommend that behind the barriers to the sea the water should be treated with wary respect. Flood defences and dikes have their place, but only among a panoply of techniques designed to work with water's natural inclinations to flow downwards, with its power for creation and destruction.

Improving soil seepage capacity will not make tourists' jaws drop, as will the storm surge barriers of the Eastern Schelde, but it will reduce discharge peaks and thus the risk of flood. The same applies to measures to maintain or enhance the runoff capacities of rivers and to reducing river flow velocity by 're-naturation' of streams. Other measures proposed include:

- prolonging warning periods by improved forecasting;
- ensuring that flood danger is taken into account in spatial planning and building projects;
- preserving and restoring wetlands and thus reducing discharge peaks;
- examining agricultural methods to the same end and in order to reduce soil erosion;
- encouraging natural forest development and afforestation to the same end;
- where possible, opening areas to flooding.[29]

The Delta Commission is still hard at work and the latest response to the threat of flood in the Netherlands came in the form of its 'advice note' to the government issued in the summer of 2008.[30] The Commission was asked by the government to give its advice on the protection of the Netherlands from the consequences of climate change, including in the extremely long term. For obvious reasons, the Delta Commission's first priority was flood protection, though it also pointed out that broad issues were involved and that climate change would touch every aspect of life.

The title of the report, 'Working with water', demonstrates the direction of the Commission's thinking. 'Safety and sustainability are the two pillars of the strategy for the coming centuries,' it asserts. 'Water safety' or 'water security' is stressed, in two broad senses: protection against floods and maintaining the safety and security of fresh water supplies. The report also states baldly that 'the level of safety must be higher than the current level by at least a factor of ten'.[31]

Its advice, said the Commission, is urgent. Existing requirements were not only inadequate, but even these flawed standards had not been properly met. The climate was rapidly changing; moreover, the sea level was rising more quickly than had been assumed and extreme variations in river runoff were increasing. This was not only increasing the danger of flood, but was threatening other climate-related catastrophes:

> A rising sea level, declining river runoffs in the summer, lengthy periods of drought and penetration of salt water via rivers and groundwater are placing the country's water supply under pressure. This is leading to damaging consequences for drinking water provision, agriculture, water transport and ... water-related economic sectors.[32]

Some of the Commission's recommendations are specific to particular regional problems but many would be relevant to any flood-threatened area anywhere in the world:

- the dikes must be strengthened and made more secure, 'improved by a factor of ten' by 2050;
- no new building in 'physically unfavourable locations', to be determined by cost-benefit analysis;
- new developments must not hinder the runoff capacity of rivers or the future capacity or morphology of lakes;
- 'build with nature' – places where sand accumulates must be preserved and research conducted into how the volume of sand accumulating can be increased;
- programmes designed to give rivers more space should be implemented in coordination with neighbouring countries;
- coordination, administration, financing and the legal framework must all be improved.

The Delta Commission report is, however, far from perfect, showing once again a tendency to compromise with short-term commercial interests wholly inappropriate in a situation in which the very survival of the nation is at stake. Paulus Jansen, opposition Socialist Party spokesman on water policy[33] broadly supported the Commission's conclusions, but nevertheless expressed a number of criticisms. The report, Jansen says, does not do enough to counter the prioritisation of the short-term, private interests which it criticises. 'In recent times places have been extensively built on which, from the point of view of water management and water safety are completely unsuitable.' This happened because local authorities 'always go for short term goals, such as extra income from land, more employment or more inhabitants, even if this causes longer term problems'. The only solution, Jansen feels, is to strengthen central government authority over land use and planning.

Jansen is also unhappy with the Commission's conclusion that the shipping route into Antwerp must not only remain fully open but must be enhanced by deepening, which he sees as an example of precisely the kind of thinking, directed at short-term gain rather than long-term considerations vital to the nation's future, which the Delta Commission purports to reject. 'Through repeated deepening of the channel,' he says, 'in order to make the inland harbours of Antwerp accessible to ever bigger ships, risks to safety and security are being heightened and silt is disappearing rapidly, an effect which will be increased by the rising sea level.' The Socialist Party argues that further deepening is unacceptable as it will increase the risk of flooding. Improved cooperation between the Belgian ports of Antwerp and Zeebrugge, and Vlissingen and Rotterdam on the Dutch side of the border would be sufficient to protect the commercial and national interests involved without putting the nation's flood security in jeopardy.[34]

A number of climatologists, on the other hand, accused the Delta Commission of having misused forecasts in order to strengthen the case for their recommendations, presenting what had clearly been 'worst case scenarios' as probabilities. The Commission vigorously defended its approach, however, and it is hard not to sympathise with the decision to base recommendations on the most pessimistic possibilities, given that an underestimate could mean catastrophe on an unprecedented scale. One Commission member, Pavel Kabat of Wageningen University, said that they had simply calculated 'the plausible upper limit'. Their intention, he said, was that this upper limit would be 'taken into account. It's just like the building of a bridge. You design it on the basis of the heaviest possible load.'[35]

Dams

Just as with dikes and other kinds of physical barriers, dams have a mixed record when it comes to flood control. The basic theory of using dams for this purpose is that flood waters can be stored behind large dams and released gradually, so that the timing of the flood peak can be both delayed and determined, while the peak itself can be effectively lowered. They can also be used to avoid the problem referred to above, of tributaries all flooding at the same time and thus arriving in the 'main stem', the river into which they all run, simultaneously.[36] The World Commission on Dams (WCD) offers a number of examples of floods having been reduced in height, with a consequent reduction in lives lost and economic damage, both by dams specifically built with flood control as a primary purpose and those originally built for other purposes: the Tarbela dam on the Indus and those of the Glomma and Laagen basins in Norway have, for example, each achieved a 29 per cent reduction in peak levels. When, during the 1995 monsoon, the Ngam Ngum River in Laos experienced three 50-year floods in a single season, the reservoir completely absorbed the first two and reduced the peak of the third by 20 per cent. Hunt also cites the contribution of the dam on the Damodar River in Bihar and West Bengal in India to the reduction of damage during the floods of 1978 and a similar event on the Yellow river in China in 1981.[37] Even when a dam can do no more than delay a flood, time to issue warnings and organise evacuations or palliative measures can save lives, livelihoods and property. The WCD notes, for example, that this has occurred in Japan, where large dams 'have dramatically reduced the sudden arrival of floods in populated areas where the rivers and exceptionally steep and short, and susceptible to flash floods'.[38]

These successes are, however, outweighed by the problems provoked by dams, especially large dams. Whether built with flood control in mind or not, most dams have actually made matters worse in two ways. Firstly, as the WCD notes, dam breaks, though rare, 'have and do occur (*sic*) and usually during exceptional storms; when they do, communities downstream are subject to extreme floods amplified by the dam break'. When the Vaiont Dam in the Italian Alps was hit by a landslide in 1964, huge amounts of water spilled over the top of the structure and 3000 people lost their lives. This was dwarfed eleven years later when two dams in China collapsed after heavy rains, killing 230,000.[39] In addition, mistakes or failings in operation, mechanical failures and extreme weather events can lead to disaster, particularly where, as the WCD puts it 'communities have adapted to the level of protection normally provided and the contingency plans – or their implementation – have been inadequate'. A further reason may be that when a dam is not specifically built for flood control, the danger of flood may be passed over or underestimated, such as when peaking operations of power stations have caused an unexpected surge of water in the river. Where safety relies on prompt warning of local populations, moreover, failure to follow procedures can also lead to catastrophe. The WCD cites a case in Nigeria where a delay in warning inhabitants led to widespread destruction of property and over 1000 deaths.[40]

Many problems stem from the fact that dams are rarely if ever built for the sole purpose of flood control, which is generally needed on no more than a few days each year. There is therefore clear scope for conflict between flood control and other objectives, such as power generation or irrigation. In general, flood control works best where reservoirs are kept low until needed. Hydropower and irrigation have precisely the opposite priority.[41] Moreover, as with other flood control measures which rely on being able to estimate flood peaks, their height and timing, climate change plays a role in making such predictions increasingly unreliable, as is also the case for changes in land use such as those discussed above when considering levees.[42] Also in both cases infrastructure is ageing just as the pace of these climate and land-use related changes is accelerating.[43]

Dams are also expensive. According to the WCD average cost overruns are 56 per cent and not only that: dams often fail to come up to expectations. Apart from their record of failure as a method of flood control, those intended for hydroelectric generation, urban water supply or irrigation are little better, with in each case high proportions falling well short of the projections on the basis of which investment was made.[44]

As public money in one form or another goes into the building and maintenance of almost every dam in the world this is a waste of resources, funds which could undoubtedly have been used to better effect. Resources are also wasted by the enormous amount of land given over to dams. The WCD found that at least 80 million people have been displaced by dams, which as well as failing to deliver on projected benefits have also created ecological problems including flooding, as we have seen, but also the salination of previously fertile land and the eroding of river banks, deltas and coastlines through arresting the flow of silt downstream. In addition to the people displaced, farming and fishing communities downstream of the dam bear the brunt of its true cost.[45]

It might be pointed out that in a world run for people rather than profit, a report such as that eventually produced by the World Commission on Dams would have been thought necessary before huge resources were committed to their construction, rather than after. In fact, in the world in which we live big dams continue to be built, including in EU member states, despite the clear evidence that the only one of their objectives they consistently achieve is to aggrandise and line the pockets of the decision-takers responsible for them. It is not as if alternatives do not exist, for flood control as well as for all of their other purposes.

The WCD's conclusion is clear. Its report lists four broad reasons why the whole concept of flood 'control' in this sense is flawed, rather than there simply being some error in the methods chosen. Firstly, 'dams have encouraged settlement in areas that are still subject to floods that exceed the maximum design flood'. In other words, sometimes due to changes in local or upstream land or river use, sometimes due to climate change and sometimes to a combination of these and other factors, the dams are simply not high enough to cope with revised flood peaks. Secondly, the cost of putting this right, of ensuring complete protection against all floods, using either dams or other structural measures, would be unacceptably high. Thirdly, the effectiveness of such measures diminishes over time. The WCD report states that the reason for this is 'the accumulation of sediment in river beds and reservoirs', but it would surely be the case anyway as maintenance costs increase with the age of the structures involved. Finally, the WCD recognises that floods are often beneficial, meaning that 'the elimination or reduction of natural flooding has led to the loss of important downstream ecosystem functions, as well as loss of livelihood for flood-dependent communities'.[46]

The WCD goes on to put some of its previous examples of positive effects into perspective. The Ngam Ngum dam, for example, was later 'blamed for a major inundation of downstream agricultural areas', becoming, in a sense, the victim of its own success 'since the reservoir had not spilled for many years people had developed a false sense of security and drains were not maintained'.[47] When the flood came this hindered the dispersal of the floodwaters and led to the total destruction of standing crops. In Poland in 1997 a flood only half as large as one which occurred in 1934 caused massively greater damage due to economic development occurring in an area supposedly protected by a dam.[48]

Adaptation and the Non-structural Approach

There is now, as we have seen, a consensus amongst the scientific, engineering and other relevant experts that the future of humanity's relationship with floods lies in adaptation and non-structural approaches to damage reduction, rather than in megaprojects which attempt to control what cannot be controlled. Crop choice, building adaptation and, most importantly, the location of any new development can all reduce the nuisance from floods and enable farmers to take advantage of their ability to fertilise land. Avoiding inadvisable use of flood plains is the only way, in the long term, to reduce the loss of life, livelihood and property which flooding means when it is combined with ill-preparedness. Hunt lists four advantages of the approach which she describes as 'Reducing flood damage by altering human behaviour rather than by altering freshwater ecosystems.'[49]

Firstly, because natural hydrological processes are not interfered with, the useful products of rivers – principally water itself, fish and other food products, and means of transport for people and goods – continue to be available. Secondly, it avoids risking making matters worse. Often, physical structures intended to protect against flood turn out to exacerbate it when it arrives. Thirdly, non-structural strategies are cheaper. And finally, they are 'adaptable to changing conditions'.[50]

These 'non-structural flood damage reduction' procedures include, first and foremost, conducting a great deal of research into 'the physical, biological and chemical processes that affect flood hazards and the natural functions of floodplains' as well as 'the social processes involved in human interaction with flood hazards and natural flood plain functions...'. Such information will include 'short-term weather forecasts, detailed stream flow data... long-term climate conditions and topographical details' as well as historical data on floods, gathered from official sources and from local knowledge.[51]

All of this information should then be brought to bear on planning decisions for the development of disaster preparedness plans, but also for the placing and design of new development, services and the whole range of facilities which human communities need or like to have at their disposal. Early warning systems developed over the last half-century have already led to substantial reductions in loss of life during floods. These systems tend to be dependent on sophisticated technologies, so that effective development aid for flood-threatened regions should include the transfer of such technologies. Their effectiveness, moreover, depends on ensuring that all affected communities and individuals are kept informed and that they are inclined to respond appropriately when so informed. Hunt cites the past experience of the inhabitants of 'squatter settlements and refugee camps' which may make them distrustful of authority.[52]

The real key to flood damage reduction is, however, not simply an efficient warning system or even sound planning regulation, but the restoration of the natural hydrological characteristics of floodplains and watersheds, coupled with the development or revival of systems of agriculture and other forms of production compatible with whatever level of flooding is unavoidable.[53]

As all experts recognise that this is the case, why are dams still being built and structural methods of flood control still being employed? One reason is simply the profit motive. Non-structural approaches to flood control by definition rarely involve major engineering projects capable of generating profit for building firms, prizes for architects or prestige for political decision-makers. Changes in land use, for example, can be expensive, but that expense does not translate obviously into profit. Once again, the problem of flood control demonstrates that, at a time of dangerous uncertainty in the face of a rapidly-changing climate, everything possible must be done to reduce the involvement of the 'market' in the conservation, organisation and delivery of water. Only detailed planning based on scientific analysis and a long-term perspective is capable of ensuring that water is available when and where we need it and that it does not appear in unmanageable quantities in inappropriate places at inappropriate times. And only the public sector, and social ownership, are capable of creating the conditions under which such planning can be carried out.

3

CONFLICT AND COOPERATION

'You steal the water from the valley, ruin the grazing, starve my livestock. Who's paying you to do that, Mr. Mulwray? That's what I want to know!'
Sheep farmer in the film *Chinatown*[1]

The nature of rivers and lakes means that conflict over water, and especially the supply of fresh water, has been recurrent throughout history on the local, regional, national and international level. Water supplies, whether from river basins, lakes or groundwater sources, which are shared by two or more countries cover almost half of the Earth's land surface and 40 per cent of its population. With the break up of the Soviet Union and Yugoslavia, the number of such shared sources has increased – from 214 in 1978 to 263 in 2006.[2] Lakes straddle and sometimes determine boundaries between cultures and nations. On their banks stand towns and rural areas whose needs may clash. The same is true of rivers, only more so. If China dams the Mekong, the South East Asian countries downriver – Laos, Vietnam and Cambodia – may lose water supply. If the United States pollutes Lake Superior, Canada gets a polluted lake too. Changes in land use on one side of the Rhine or the Rio Grande will have an impact on land use on the other side and so on.

Water has been the cause of numerous border conflicts. These can result not only from the kinds of inherent tensions mentioned above, but water itself can also be used as an instrument of power. One group of people – a community, a nation, an armed gang or a corporation, for example – uses its control of water to impede everyone else's access and bully out of them whatever it is they want. As the OECD notes in its study of water as a source of conflict:

As water scarcity can affect different groups in different ways, tensions may arise among them. The risk is particularly high where there is discrimination over access to water. Water may at times even be used as a 'weapon' or tool for oppression against a marginalized group.[3]

What applies to a society may also apply to relations between states. Examples include Israel's use of its control of the Jordan River against its Palestinian and other Arab neighbours,[4] South Africa's threat to withhold water as a means of controlling its black population under apartheid and US attempts to exert control over the Guarani Aquifer, which supplies water to four South American countries. The apparent use of irrigation systems in much of the ancient world to establish and maintain authoritarian political systems is a paradigm of ancient history.[5]

Water can generate tensions in three discrete ways. Firstly, where a supply is finite and must be shared, whether between individuals, communities, economic sectors, regions or nations, conflict is likely. The HDR for 2006 gives the example of a situation in which 'the retention of water upstream for irrigation or power generation in one country restricts flows downstream for farmers and the environment'. Secondly, the way an upstream country or community uses water may affect its quality downstream. This may involve dam development interfering with sediment deposits, industrial, agricultural or urban pollution from ongoing activities which may change with development, or one-off events such as a chemical spill. Finally, irrigation projects, dams or flood control measures may affect the size and timing of water flows downstream.[6]

So, as more and more countries experience shortages of water, its role as a provoker of conflict becomes more acute. This is not, however, a simple matter of mathematics or logic, of simply observing that the shorter a resource, the more people are likely to fight over it. We are not talking about thirsty communities driven to desperation, but of demagogues and greedy corporations eager to exploit such desperation for their own ends. It is the growing treatment of water as a commodity, a tradable good, which is sharpening the potential for conflict.

Anil Naidoo, director of the Blue Planet Project, a group which campaigns for the recognition of access to clean water as a human right, notes that growing 'pressure on freshwater resources' is leading to 'tangible consequences for human communities as well as on ecosystems' and that these reflect what he calls 'further forms of structural violence'. Naidoo defines 'structural violence' as a phenomenon which stands in a mutually reinforcing relationship with 'direct or personal violence represented by a lack of access to safe water for basic sustenance and sanitation' and which consists of 'policies and political influence that advocate that water be defined as a commodity'. This can lead to inter-state, inter-regional and inter-communal conflict because it creates a context in which 'concerns of geopolitical positioning and power are inseparably intermeshed with concerns of equity of access and market-based agendas'.[7]

The usual methods of avoiding armed conflict – effective international treaties based on mutual self-interest and give-and-take compromise, for example – are as available for resolving international disputes over water as they are in other instances. Yet the fatal synergy of climate change and neo-liberalism is making it increasingly difficult to arrive at such agreements. Many researchers nevertheless strike a cautious note of optimism over the role of water in fostering good neighbourly relations. The University of Oregon's 'Basins at Risk' (BAR) research programme, for example, which sought on the basis of an historical analysis to identify risk factors potentially leading to dangerous conflict, found that it was often the case that 'countries exhibit greater cooperation over water than overall, indicating that countries in conflict over other concerns may still find common interest in cooperation with regard to their shared water resources'.[8]

The BAR researchers also found that there was little or no evidence in support of the 'commonly cited indicators linking freshwater to conflict', listing 'spatial proximity, government type, climate, basin water stress, dams...development...' and 'dependence on freshwater resources in terms of agricultural or energy needs'. International treaties were shown by BAR to have been an effective mechanism for preventing or resolving conflict.[9]

Indeed, the Permanent Indus Water Commission oversees a treaty which includes a mechanism for dispute resolution so effective that it continued to function through two wars between the two parties to the treaty establishing it, India and Pakistan.[10]

Arun P. Elhance, a US geographer, has conducted a series of case studies of international river basins and concluded that 'the hydrology and geography' of such a basin 'tie all the riparian states sharing it into a highly complex web of interdependencies, leaving them no choice but to interact with one another indefinitely'. Moreover:

> These interdependencies grow with time as the demand for water for multiple needs grows in all the riparian states. Although states are inherently inclined to unilaterally exploit the rivers flowing across or along their borders, the hydrologically induced interdependencies in international basins gradually compel states to entertain at least the possibility of cooperation with their neighbors.[11]

Not everyone shares this optimism. The French Catholic development organisation Comité Catholique contre la Faim et pour le Développement (CCFD) lists several international conflicts which either have already involved armed clashes or which it believes could lead to war:

- over the Senegal river, between Senegal and Mauritania, which has been the subject of armed skirmishes in the relatively recent past;
- over the Nile, between Sudan and Egypt;
- over the 'megadams' on the Parana river, between Brazil and Argentina;
- the war in Kashmir, an issue in which is the division of water between India and Pakistan;
- over the Ganges, between Bangladesh and India;
- over the Mekong, between China and the downriver states;
- the Israel–Palestine conflict, in which access to the River Jordan plays a role;

- between Turkey on the one hand and Syria and Iraq on the other, with the two Arab states accusing their northern neighbour of 'exhausting their reserves of water' by constructing dams on the Euphrates and Tigris.[12]

Other conflicts might be added to this list: United States–Mexico over the Colorado River, whose waters are so heavily depleted that they reach the sea only intermittently; Iran–Iraq; Namibia–Botswana–Lesotho–South Africa and Uzbekistan–Kazakhstan–Kyrgyzstan–Tajikistan.[13] Hungary–Slovenia, on the other hand, who have long disputed the waters of the region, appear to have settled their differences amicably but to have seen the results spill over into closer cooperation on a range of issues. This was evident even before accession to the European Union, or the establishment of the Water Framework Directive, with the establishment in 1995 of the Hungarian–Slovenian Water-Management Committee and provides an example of the way in which shared water resources can sometimes bring countries closer together, as well as creating or exacerbating conflict.[14]

Water Wars?

As a World Bank paper on water policy points out, while it is

> reasonably argued that there has rarely been a 'water war' ...it is probably the case that there has never been a single cause for any war, and resource conflicts – land, water, minerals – are clearly common contributory factors to many past and present (and future) conflicts.

We would go further than this and argue that at root all wars are conflicts over resources. As water is the most important of all resources, it is likely that wars will be and have been fought in part, at least, over it. As the World Bank paper continues, '... water plays a significant part in a number of recent and current disputes, even conflicts, around the world, especially where climate variability, coupled with major transboundary flows, create high levels of perceived threats to national water security'. However, for precisely the same reasons '...cooperation with regard to shared waters contributes to strengthening relations between countries, and catalyzing broader cooperation, integration and stability'.[15] Often, whether states are able effectively to cooperate or are forced into conflict will depend on dynamics within their societies, dynamics within which the profit motive can only be a destabilising influence.

Antonia Juhasz, a senior analyst at the progressive US periodical *Foreign Policy in Focus*, joins the World Bank experts in pointing out that not all international conflict over water expresses itself in a straightforward way in which Country X confronts Country Y over access to a certain water supply. Water, she notes, has been treated by the US invaders as part of the spoils of the Iraq war, with US transnational corporation Bechtel profiting from a $608 million contract that gives it access to Iraq's water and potentially that of much of the Middle East. Bechtel, one of the world's leading water privatisers, will be responsible for rebuilding Iraq's water and wastewater systems, destroyed, incidentally, in a war for which the company was an enthusiastic advocate. Bechtel already has interests in more than 200 water and wastewater treatment plants around the world. Juhasz gives some examples of what this means for those who have to rely on the company to provide these services:

> after privatizing the water systems in Cochabamba, Bolivia, a subsidiary of Bechtel made water so expensive that many were forced to do without. The government met public protests with deadly force. Bechtel waited. Finally, the government canceled Bechtel's contract. Bechtel responded with a $25 million lawsuit for lost profits.[16]

In an 'Issues Brief' entitled 'Water and Violent Conflict' published in 2005 by the OECD, the club of developed nations expressed its concern over 'water-related tensions' and noted that 'access to water and water allocation and use can become the focus of tensions, which may potentially spill over into conflict, with or between states'.

In common with others on all sides, however, the OECD felt that violent conflict between nations was unlikely, that 'violent conflicts over water are most likely on a local level, for example, *over the privatisation of drinking water* or access to a water point' (authors' emphasis).

The OECD also listed a number of aspects of any possible conflict over water, both positive and negative: '...tensions between countries that share a river basin may hinder sustainable development – thus indirectly driving poverty, migration and social instability. They also have the potential to exacerbate other non-water-related violent conflicts.' It identified 'four interdependent levels' on which violent conflict could emerge between and within states:

- the local level: for example, between societal groups over access to a water point; or between the state and people affected by the construction of a dam;
- the national level: for example over the reallocation of water between economic sectors;
- the international level: between upstream and downstream states;
- the global level: for example, between food exporters and food importers in relation to the world food market.

However, water may also help to resolve conflicts, as a result of these factors:

- a mutual need to share water may be used to help forge peaceful co-operation between societal groups;
- between states, the development of shared data, information systems, water management institutions and legal frameworks helps to sustain efforts to reduce the risk of conflict.

These different levels and phenomena are of course not independent: 'any intervention affecting one level needs to assess the impacts on the other levels', which include both physical links (water, food) and non-physical links (political processes, economic flows).[17]

There is a surprisingly wide consensus when it comes to conflict over water, and it is that you are much more likely to come to blows with the people from the next town, or the next street, or the next house over immediate access than you are to see your sons (and, in our times, perhaps you daughters too) marched off to fight wars over it. The OECD again:

> Millions of the world's poor, particularly in rural areas in subsistence agricultures, depend on water for their livelihoods. If no adequate economic and political measures are taken, water will contribute to instability in water-scarce regions as demand approaches the limits of supply...Water scarcity, in combination with low economic development and shortfalls in political governance and other mechanisms in managing tensions peacefully, may therefore lead to instability – threatening lives and livelihoods.[18]

Although such conflicts seem inherently local in nature, the tensions they create can easily be exploited by people pursuing power, or can for other reasons spill over borders and become international conflicts, as we have seen amongst resource-starved people in Africa.

Public Ownership

The OECD's view – one which is widely shared, and shared by the authors – is that it is not in general the absolute level of availability of water which determines whether it is likely to become a contested resource. Whether water emerges as a source of conflict depends on the way in which it is governed, administered and distributed. 'Whether scarce or not,' the OECD argues, 'the highly complex and sensitive nature of its availability, use and allocation requires strong, capable mechanisms and institutions to negotiate and balance competing interests and to manage this vital resource.' This is 'at least as important as traditionally cited variables such as climate, physical water availability, population density, and levels of economic development'.[19]

Where the authors would differ from the OECD though is that we regard this as a cast-iron argument for public ownership. If sound and impartial administration of water supplies and sanitation are to be achieved, then where water is in public ownership it must be kept in public ownership, and any problems of inefficiency, corruption or discrimination dealt with within that framework. Where it is not in public ownership, one of the competing interests can be removed by socialisation of the resource, by taking it into state or municipal ownership and establishing meaningful structures enabling citizens – emphatically not 'stakeholders' – to participate in decision-making.

Private companies have no place in the water supply industry. Demand-side management, which the OECD correctly argues is 'the only sustainable form of water management' and which includes 'reuse, more efficient technology, institutions to safeguard sustainability', but also dealing with leakage and other sources of waste, can only be organised on public sector principles. It is the job of a privatised water company to sell water, just as it is the job of whoever makes computer games to sell computer games. The only real responsibility of a private corporation is to make profit, to earn money for its shareholders. The more water it sells, the more money it will make. Even dealing with waste must be done according to the principles of profit-targeted investment. Is it worth investing X in improving this conduit when it will save only Y per year? But this is, of course, X dollars and Y dollars, not X dollars and Y litres of water, a real rather than a symbolic resource. One of the many things which makes water unique is the huge gap between use value and exchange value. The former could hardly be higher, yet apart from in the most exceptional circumstances (usually involving the exploitation of the poor and vulnerable), the latter is low. And there are two other features which separate water from computer games: in all but rare and exceptional cases, water is a natural monopoly; and when the computer games run out, people will move on to some other kind of game and nobody will die.

Every one of the OECD's criteria for good governance can, in our opinion, only be achieved within a public ownership model. Firstly, the OECD insists on 'stakeholder dialogue and participation'. We prefer to speak of popular democratic control, 'stakeholder' being too readily interpreted to include corporate water suppliers and users. In our view, a corporation active in an area and using its water has certainly a right to express its opinions in open fora, but those who own or control it have only the same rights as other individual citizens. The task of democracy is not to facilitate the expression of the corporate will but to restrain it. Corporations are powerful entities, potentially able to determine such crucial questions as whether people work, how land is used, how natural resources are exploited. One of the most important tasks of democratic government in the particular circumstances of the early twenty-first century is to enable the people to resist such power so that decisions may be taken for the good of the community as a whole. The 'stakeholder' concept attempts to circumvent such reasoning by presenting the will of a corporation as privileged, while at the same time appearing to be committed to public participation in decision-making.

Secondly, the OECD correctly notes that 'a lack of clear responsibilities between (state/public) institutions may lead to failures in effectively and peacefully managing competing claims and practices'. Again, the introduction of a gain-seeking third party into such a potentially delicate balancing of tasks seems unlikely to be helpful. Corporate capital invariably brings with it its own traditions of ownership, which, having been forged in the particular conditions of early modern Europe and later North America, will often fail to mesh with local custom and practice. The state is much more likely to be in a position to treat such matters with sensitivity. Even if this has often not been the case, removing this responsibility from the state seems neither wise, nor, in the end, possible. Finding ways to pressure states into fulfilling it, of performing the task of mediating competing interests within their jurisdiction, is a worthy goal, and one which can only be complicated by the presence of corporations as profit-making service providers.

As for the third OECD criterion, the managing of internationally shared water systems, again this can clearly only be fulfilled by states, and public ownership of water resources can only facilitate this task. 'Complex physical, political, and human interactions can make the management of shared water systems especially difficult.' To add a private corporation to this mix is, once more, to complicate it unnecessarily, to bring into the mix an ethos which is likely to be quite alien to those which have had to be mediated in the past.[20]

States have generally a good record in resolving conflicts over water, or at least of ensuring that they do not become violent conflicts. Since the end of the Second World War almost 300 international agreements have been signed governing water supply, water and economic development and water as a source of energy. Blood has been spilled in conflicts in which water played a role, but the vast majority of this has been to ensure corporate control of water in Iraq. The only state which has, since the Second World War, seized water resources belonging to its neighbours is Israel'[21] whose belligerence is made possible by the same global power as is responsible for the invasion of Iraq. Thus, two colossal acts of international criminality have been committed or colluded in by a country in which it is impossible to become President, and difficult to hold any other position of influence, without corporate backing. In our century, whenever states resort to aggressive violence, corporate interests are invariably at work.

Finally, the OECD advocates the establishment of a reliable database of meteorological, hydrological, and socio-economic data as a means of ensuring 'effective longer-term water resource management, particularly across regions'. It notes that:

> Tensions between states can emerge when data is not shared appropriately or is misused to block development plans. Disparities among water users' capacities to generate, interpret and legitimize data can lead to mistrust, hindering co-operation. However, collaboration and trust can be built up by sharing data transparently and engaging in open discussion.[22]

Two things which do not characterise the behaviour of corporations, simply because they are rarely if ever in their interests, are 'sharing data transparently and engaging in open discussion'. Under whatever ownership model, this will be the responsibility of states. Again, separating ownership from such a crucial aspect of the problem seems ill-advised.

Benefits of Cooperation

States will cooperate with each other rather than entering into conflict when it is in their interests to do so. The very quality of water which makes it such a ready source of international conflict – the fact that, unlike other resources, it flows across borders – has also made it a source of frequent cooperation. Two World Bank water specialists, Claudia W. Sadoff and David Grey, have suggested a useful taxonomy of benefits derived from rivers which helps explain why this is so. First, they say, come 'benefits to the river': by working together, states can 'enable better management of ecosystems, providing, and underpinning all other benefits that can be derived'. Secondly, 'benefits from the river' will be enhanced by such management, and will include 'efficient, cooperative management and development' which can in turn result in huge benefits, including increased food and energy production. Thirdly, benefits from the river: as such cooperation can reduce tensions, and 'as tensions between co-riparian states will always be present, to a greater or lesser extent, and those tensions will generate costs', such costs will be reduced. Fourthly, benefits from the river: 'as international rivers can be catalytic agents, cooperation that yields benefits from the river and reduces costs because of the river can pave the way to much greater cooperation between states, even economic integration among states, generating benefits beyond the river'.[23]

This final category of benefit is of crucial importance to the argument for public ownership, involving as it does benefits to the broader society which could logically have no place in the investment arithmetic of a private corporation. As Sadoff and Grey argue, 'cooperation in the management and development of international rivers may contribute to, or even result in, political processes and institutional capacities that themselves open the door to other collective actions, enabling cross-border cooperation *beyond the river*' (emphasis in original).[24] In other words, planned use of water resources stands at the very centre of economic development, both on the national and international level, and for that reason, amongst others, is absolutely unfitted to be surrendered by the state to the forces of the profit-driven market.

Trans-boundary management is clearly possible, given that neighbours have so often been able to come to mutually satisfactory arrangements over water. Where water must be shared between nations exploiting the same source, the United Nations Development Programme would like to see it 'manage(d) in an integrated way across the whole basin, with countries trading agricultural resources, hydropower and other services according to their comparative advantage in water use'. So, for example, because 'hydropower is more cost-effective in sloping mountainous upper reaches, while irrigation produces better results in valleys and plains...trading hydropower for agricultural goods is one way of tapping into this comparative advantage'.[25] Sadoff and Grey give examples, moreover, of instances in which cooperation which has begun with agreement over water resources has developed much further. In South East Asia, they argue, sharing the resources provided by the Mekong 'has proved to be an important stabilizing factor....bringing substantial benefits "beyond the river", both directly from forward linkages and indirectly from diminishing tensions':

> During years of conflict between Laos and Thailand...Laos always provided hydroelectricity to Thailand, and Thailand always paid. Similarly, the government of Thailand has followed an explicit strategy of increasing regional stability by creating mutual dependency and thus purchases gas from Myanmar and Malaysia and hydropower from Laos and China...[26]

Bilateral cooperation has in general proved much easier to achieve than the sort of multilateral cooperation demanded by rivers such as the Mekong which pass through more than two states, however.[27] With each riparian state, complexities increase exponentially. Cooperation both on the immediate issues generated by river resources and on these 'beyond the river' matters thus requires sophisticated institutional mechanisms for establishing procedures and resolving disputes, the creation of which can only be complicated by the involvement of transnational corporations with their own agendas.

Treaties and Agreements

The problem for international cooperation over water is that under normal circumstances states have an absolute sovereign right to use resources found on their territory. In the case of water, however, the rights of one state will clearly clash with another where water resources are shared. Absolute sovereignty is sometimes challenged by downstream states on the alternative legal principle of 'prior appropriation', a sort of 'grandfather clause' under which past use establishes a right to future use.[28]

The first attempt to go beyond this was taken in 1911, with the adoption of the Madrid Declaration on the International Regulation regarding the Use of International Watercourses for Purposes other than Navigation. This was not an international treaty but merely a set of guidelines approved by the Institute of International Law. Nevertheless, the Madrid Declaration had sufficient resonance to be seen as of continuing relevance 55 years later, when the 1966 Helsinki Rules on the Uses of the Waters of International Rivers incorporated most of its points whilst at the same time superseding it.[29]

Although they have never been adopted by the United Nations and are therefore non-binding, the Helsinki Rules have been extremely influential and are generally regarded as prevailing in all situations in which no bilateral or multilateral treaty on shared water overrides them. The basic principle underlying the Rules is that each lakeside or riparian state should be accorded a 'reasonable and equitable share in the beneficial use of the water' and that states should avoid actions which lead to 'appreciable harm' to other riparian states. As is so often the case with international agreements, whether binding or not, much is left to later statespeople, legal experts and diplomats, especially when it comes to agreeing definitions of such slippery concepts as 'reasonable', 'equitable' and 'appreciable harm'.[30]

The Helsinki Rules have been further developed, for example by the inclusion of groundwater in 1989. In addition, many bilateral and multilateral treaties have been signed and from these, as well as the refinements of the Helsinki Rules themselves, one can derive certain common factors defining the concepts which the Helsinki Rules left to be settled by later negotiations. What must be taken into account when determining what is 'reasonable and equitable' and what constitutes 'appreciable harm' are the natural attributes of that portion of a watershed within a state; the prior or existing use of the water; the social and economic needs of the population; alternative resources to serve those needs, and the comparative cost of these; and the need to avoid damage to other riparian states or communities, especially downstream users.[31]

The next significant development in international water governance was the International Conference on Water and the Environment held in Dublin in January 1992. This conference of experts, rather than politicians or diplomats, issued what were called 'The Guiding Principles from the Dublin Statement (1993) on water and sustainable development', which were as follows:

- Principle No. 1: Fresh water is a finite and vulnerable resource, essential to sustain life, development, and the environment. Because water sustains life, effective management of water resources demands a holistic approach, linking social and economic development with protection of natural ecosystems. Effective management links land and water uses across the whole of a catchment area or groundwater aquifer.
- Principle No. 2: Water development and management should be based on a participatory approach, involving users, planners, and policymakers at all levels. The participatory approach involves raising awareness of the importance of water among policymakers and the general public. It means that decisions are taken at the lowest appropriate level, with full public consultation and involvement of users in the planning and implementation of water projects.
- Principle No. 3: Women play a central part in the provision, management, and safeguarding of water. This pivotal role of women as providers and users of water and guardians of the living environment has seldom been reflected in institutional arrangements for the development and management of water resources. Acceptance and implementation of this principle

require positive policies to address women's specific needs and to equip and empower women to participate at all levels in water-resources programs, including decision-making and implementation, in ways defined by them.

• Principle No. 4: Water has an economic value in all its competing uses and should be recognized as an economic good. Within this principle, it is vital to recognize first the basic right of all human beings to have access to clean water and sanitation at an affordable price. Past failure to recognize the economic value of water has led to wasteful and environmentally damaging uses of the resource. Managing water as an economic good is an important way of achieving efficient and equitable use and of encouraging conservation and protection of water resources.[32]

These principles, though overlaid with political and diplomatic compromise, can be discerned in the 1997 UN Convention for the Non-Navigational Use of Shared Watercourses, which added to the Helsinki Rules the important 'prior notification of works' requirement. The broad idea of the 1997 UN Convention, according to the 2006 HDR, is that

> governance of international watercourses should be developed by taking into account the effects of use on other countries, the availability of alternative water sources, the size of the population affected, the social and economic needs of the watercourse states concerned, and the conservation, protection and development of the watercourse itself.[33]

Unfortunately, the Convention provides no mechanisms for interpreting these concepts, or for resolving disputes. As a paper produced for the World Bank on the benefits of international cooperation over rivers notes, the key principles on which it is based – 'equitable utilization' and 'no significant harm' – are difficult to apply and sometimes mutually contradictory, so that upstream states are more likely to embrace the first principle and downstream states the second, whereas the reality is that 'upstream extraction generates externalities downstream by diminishing flows physically. On the other hand, downstream extraction can generate externalities upstream by diminishing future available flows upstream because of downstream claims of acquired rights to that water.'[34] Again, the Convention fails to provide effective mechanisms for resolving such competing claims.

A more promising model for international cooperation might be the United Nations Economic Commission for Europe's Convention on the Protection and Use of Transboundary Watercourses and International Lakes (ECPUTW). The UN Economic Commission for Europe (UNECE) is a regional Commission of the UN Economic and Social Council designed, in its own words, 'to promote cooperation among its member states', which include almost all European countries, the five central Asian post-Soviet 'Stans', Israel and the United States and Canada. Signed in 1992, the ECPUTW came into force in 1996 and has been ratified by 35 countries and the European Union. It also contains a mechanism enabling it to be extended to become global. This requires that 23 countries which are not member states of the UNECE ratify, and admittedly so far only four have done so.[35]

Cooperation within Europe may be less difficult to achieve because of the relatively stable nature of most of its states and the existence of numerous bodies capable of mediating disputes – the European Union, the European Economic Area, UNECE itself, the Council of Europe and a plethora of specialised councils, association agreements, treaties, conventions and so on. On the other hand, on a small continent which has 150 transboundary rivers, 25 international lakes and more than 100 internationally-shared aquifers, the difficulties to be overcome will clearly be extensive.[36]

Its main provision, contained in Article 2, is that 'Parties shall take all appropriate measures to prevent, control and reduce any transboundary impact', such impacts being defined as 'adverse effects on the environment caused by a human activity, including effects on human health and safety, flora, fauna, soil, air, water, climate, landscape and historical monuments or other physical structures or the interaction among these factors'. The Convention aims to protect transboundary waters 'by preventing, controlling and reducing pollution' but, beyond this, seeks to develop and encourage what it calls 'integrated management' of such shared waters. As well as 'conflict prevention' it promotes the 'conservation and restoration of ecosystems'.[37]

The 'principles' upon which the Convention is purportedly based will be familiar to anyone who has followed the development of environmental policy within the European Union over the last two decades, though so will the fact that they are more readily and more frequently uttered than respected. These are: 'the precautionary principle', a moral and political principle which states that if an action or policy might cause severe or irreversible harm to the public or to the environment, in the absence of a scientific consensus that harm would not ensue, the burden of proof falls on those who would advocate taking the action,[38] 'the polluter-pays principle' and 'the principle of sustainable water management'.[39]

What distinguishes this Convention from other multilateral global or wide-regional agreements is its establishment of mechanisms designed to implement its goals, both legally binding tools and policy guidelines. These include the definition of water quality objectives, the licensing and monitoring of waste-water discharges, emission limits on discharges of hazardous substances, including that these be based on best available technologies, and minimum levels of treatment for municipal waste waters. The Convention contains a requirement for environmental impact assessment for projects affecting water courses and provisions to encourage an approach based on the ecosystem and on contingency planning, including measures to minimise the risk of spillage and other sources of accidental water pollution. Most interestingly, parties bordering the same transboundary rivers or lakes or having access to the same groundwater sources are obliged by the Convention to conclude 'specific bilateral or multilateral agreements' which should include provision for the establishment of joint bodies. The EU Water Framework Directive, which will be examined in detail in the next section, is one such agreement, and there are also conventions covering the Danube and its first and second order tributaries, the Elbe, the Meuse, the Rhine Basin, the Moselle and Sarr rivers, Lake Constance, which fronts on Austria, Germany, Liechtenstein and Switzerland, the

Rhone, Lake Geneva, the Scheldt, the Oder, the Aral Sea, the Chu Talas Basin between Kazakhstan and Kyrgyzstan and two bodies covering the North American Great Lakes. All of this is facilitated by the encouragement of capacity building through technology and know-how transfer and sharing, and the creation of a central advisory service. This advisory service, accompanied by a legal board and expert groups, is linked to a secretariat which in turn serves working groups on integrated water resources management, monitoring and assessment, and water and health.[40]

On paper, at least, the Convention clearly represents an effective instrument of international cooperation. In fact, the avoidance or peaceful settlement of water disputes within and beyond the area covered by the Convention, and the clear willingness of states to go to the negotiating table rather than the battle field when potentially conflicting interests over water are present, demonstrates that if the oft-repeated prediction that the wars of the twenty-first century will be over water turns out correct, it will be because states have lost the power to control the supply of water to their populations and to negotiate freely with their neighbours. And they can only lose this by handing the show to corporations, for whom war is merely another route to profit.

This can be seen equally when it comes to the most successful form of international cooperation concerning water, the 'basin treaty'. Basins have been as much a source of cooperation towards mutual benefit and development as a source of conflict. Yet 158 of the world's 263 international basins are not subject to any treaty or agreement which establishes any form of joint resource management, while many others are subject to agreements which do not include all of the riparian states. Treaties dealing with the increasingly crucial question of volume sharing – 'volumetric allocations', in the jargon – are in the minority, moreover, with only around a third of them addressing this issue and the rest dealing with such questions as hydroelectricity, flood and pollution control. Many treaties and cooperation agreements are quite limited in scope, dealing with a single, highly specific issue.[41]

Many of the treaties which have been concluded, moreover, lack proper structures for conflict resolution. Without such structures, crucial matters such as water allocation, equitable distribution of costs and benefits, control of pollution, management of flooding, adaptation to climate change, land-use change and other kinds of change or development become subject at best to goodwill and at worst to the law of might is right. Clearly, if step one in the successful management of water resources is for states to remain in control, step two is for them to cooperate with each other to maximise benefits.[42]

Beyond the Treaty

If cooperation is to be not only broadened, but deepened, states must first deal with a number of obstacles standing in the way. First and foremost is what the United Nations Development Programme describes as 'competing claims', including those which are seen as imperative to national sovereignty. National economic plans often take little account of the constraints imposed by the need to share water with neighbours, even when this need has been recognised in treaties with those neighbours. The result is that aggregation of the different plans reveals that they are incompatible, that together they generate unsustainable demand.

Secondly, the UNDP notes what it calls 'weak political leadership'. This is due less to the failings of political leaders as individuals, but more to the fact that they are 'accountable to domestic constituencies, not to basin-sharing communities and the governments that represent them'. In other words, 'more equitable water sharing might be good for human development in a basin, but... a vote loser at home', while 'the domestic benefits of sharing are unlikely to come on stream during the term of office of any one government'. These considerations are likely to be most immediate in parliamentary democracies, but no government can afford to ignore the views and feelings of the people it governs for long.

Thirdly, 'asymmetries of power' which 'shape the willingness to cooperate, negotiate and share benefits' can have the effect of undermining trust. Finally, where some countries within a basin refuse to cooperate, this may discourage others from doing so. China's dams threaten the Mekong River, but nothing can be done within the Mekong River Commission because China is not a member.[43]

In concluding his study of six internationally shared river basins, Arun Elhance notes that states sharing such basins

> are confronted with what seems like a simple choice: either to engage in protracted and costly conflicts with their neighbors over the rights of ownership and use of water resources, or to cooperate with their neighbors to develop and share the bounty that water in its multiple uses can help produce.[44]

Each situation is unique, however, and uniquely complex, so that

> the choices that are and can be made by states in conducting hydropolitics with their neighbors depend on the *unique* combinations of the geographical features of a given basin with a multiplicity of historical, political, economic, social, strategic, and cultural factors and circumstances specific to that basin.[45]

These complexities may stand in the way of effective international cooperation. Sharing and cooperation may be seen as endangering security, especially where there is a history of antagonism between states sharing a river basin or lake. Fear of dependency may compound this, especially among downriver states who may be afraid that their water supplies could be turned off at the whim of a neighbouring, upstream government. Such considerations may cause divisions within countries as well as between them, as may any unevenness of water distribution, whether natural or engineered. If one part of a country has sufficient water and another does not, people in the latter region are going to want to know why. This may also hamper international cooperation, with water-scarce areas of a country opposing any scheme which appears to give access to neighbouring countries before their needs are satisfied. This may sometimes be what is actually happening, but drought-stricken people may also be susceptible to manipulation by demagogues.[46]

All of these problems, as we have said, can only be made worse by the introduction of commercial interests in the form of corporations to which water is simply another commodity. There are many ways in which developed countries could be addressing the problem of water supply in poorer parts of the world, and encouraging cooperation between states sharing river basin and lake waters is one of the more potentially effective. If the European Union and the international financial and trade institutions can use their muscle to force liberalisation and privatisation programmes on to developing countries, they could equally use it to encourage cooperation between neighbours towards equitable solutions to conflicts or potential conflicts. This would be to everyone's benefit, rather than only to that of the corporations which stand to gain profits by usurping developing countries' water and selling it back to them, the real meaning of privatisation.

4

IT NEVER RAINS BUT IT POURS: CLIMATE CHANGE, WATER SHORTAGE AND FLOOD

'All across the world, in every kind of environment and region known to man, increasingly dangerous weather patterns and devastating storms are abruptly putting an end to the long-running debate over whether or not climate change is real. Not only is it real, it's here, and its effects are giving rise to a frighteningly new global phenomenon: the man-made natural disaster.'

Barack Obama

It has now been firmly established that the warming of the climate is having a number of worrying impacts on the global hydrological cycle and the systems which depend on it. Some areas are experiencing higher levels of rainfall, some prolonged drought. The frequency and intensity of extreme events are increasing. The polar icecaps, glaciers and mountain ice are melting, which affects the level and annual cycle of rivers and lakes, and the global sea level. Evaporation rates are rising, the level of water vapour in the atmosphere rising and levels of soil moisture and runoff are changing.[1]

As forests die or are altered in nature as a result of changes in precipitation patterns, the effects will reverberate through the system, affecting rivers and lakes in the vicinity of the forests and in some cases well beyond. On Kilimanjaro, for example, the forests which grow up to 3100 metres above sea level provide almost all of the water which comes off the mountain, much of which drains underground into waterholes used by people and animals, many of them far away. The higher forests are threatened by an increased incidence of forest fire. Added to a much smaller loss of runoff from the mountain's disappearing glaciers, this will lead to a catastrophically diminished water supply whose impact on the livelihoods of people dependent on the affected river, as well as on Tanzania's hydroelectric energy supplies, will range from damaging to catastrophic.[2] In Europe, while the north becomes wetter, the south is becoming dryer, with precipitation, runoff, lake levels and river discharge all dropping, bringing the threat of much more frequent and sustained drought.[3] Changes in annual river discharge are predicted to be particularly significant, decreasing by up to 50 per cent by 2070 in the south and increasing in the same period by the same proportion in the north.[4] Geographical schism will also be experienced in Australia, with the tropical north in this case expecting increased rainfall as the south becomes ever dryer. Australia's temperate zone already appears close to crossing the line from 'drought' into permanent aridity.[5] A similar situation prevails in China, where the south is likely to experience more flooding as temperatures rise, while disturbances in the pattern of monsoon winds threaten the north with drought.[6]

On a global level total annual river runoff is likely to increase, though the regional pattern is highly varied. Runoff itself may not help to avert drought in any case, unless infrastructure can be adapted to capture and store or distribute the additional water.[7] An increase in total runoff is thus more likely to mean simply an increased frequency of floods, rather than representing a mitigating factor, especially as flows during dry periods will be lower.[8] Much the same goes for one of the reasons why river flow overall will increase, the predicted continued global rise in precipitation, projections for which are 3 per cent at 1.5 degrees increase and 15 per cent at 3.5 degrees.[9] This overall increase will not help at all unless the infrastructure to harvest and store it is available in the right place at the right time, and will therefore even in the best possible scenario be of much less significance than the disruptions caused by changing rainfall patterns and other effects of global warming on the real availability of water.

As well as precipitation, river flow and water availability are affected by the rate of evapotranspiration, the process by which plants lose water through their stomata (often called 'pores') in order to regulate their temperature and maintain their life processes, a rate determined by a complex of factors including net radiation, and changes in atmospheric humidity and in the rate of air movement across evaporating surfaces. Different plants have different rates of evapotranspiration under different conditions. This means that changes in plant communities as a result of either warming or greater or lower levels of precipitation, as well as related phenomena such as extreme weather events or changes in the dependent fauna or human population, can all result in alterations to the rate of evapotranspiration. This rate determines in turn the amount of water being taken from the system and the amount being returned. Individual plants also tend to grow bigger when the proportion of carbon dioxide in the atmosphere they are breathing goes up.[10] In the far north of the Northern Hemisphere, including the far northern parts of Europe, increases in precipitation will exceed any accompanying increase in the rate of evapotranspiration, with the result that soil moisture levels will rise. However, further south in the more agriculturally productive areas of Europe, soil will become dryer due to a combination of higher temperatures and lower rainfall.[11]

Water Quality

Climate change has implications not only for the quantity of water available, but also for its quality. In common with other ecosystems, freshwater environments involve highly complex networks of relationships. An increase in water pollution can result from a number of factors related to climate change, quite independently of any change to the nature or level of polluting emissions. These may include a rise in water temperatures; an increase in the intensity of precipitation; or long periods when the flow of water in a particular river is unusually low. All of these phenomena may result in an increase in water pollution, bringing human and animal sickness and major environmental problems in their wake as algal, bacterial, fungal and other forms of infestation spread. Lakes may in some cases be subject to reduced oxygen levels and higher concentrations of phosphorus as higher temperatures change their internal thermal patterns, disturbing sediment. On the other hand, earlier ice melting in the spring may improve the quality of some lake waters. Even this, however, may not be all good news, having unpredictable effects on a lake's mix of species and thus on its own ecosystem and the broader ecosystems which surround and interact with it.[12]

Freshwater ecosystems are vulnerable to events upstream, whether direct human activities, the indirect results of them, or entirely independent of them. Apart from pollution from industry or agriculture, changes in the amount of water flowing through a river or into a lake can result in changes to the ecosystem through modifications in water levels, water chemistry, or currents. While cold water habitats will become scarcer and warm water habitats increase, the knock-on effects of these and related changes will be both much more complex and much less predictable. Biodiversity loss is not only undesirable in itself, its impacts on water quality and other aspects of the aquatic ecosystem can be cumulative and in some cases, possibly catastrophic.[13]

As numerous writers and activist organisations have been at pains to stress, when dealing with freshwater ecosystems and even more so in the case of wetlands, we are in most cases already faced with problems brought about by various forms of environmental degradation. Around 50 per cent of wetlands are estimated to have disappeared since 1900, with the second half of the twentieth century seeing the blight spread from temperate zones in Europe and North America to tropical and sub-tropical areas.[14]

Wetlands have for the most part fallen victim to economic developments involving drainage for agriculture, housing and other uses. Direct harvesting of products through, for example hunting, as well as river embankment and damming and the over-exploitation of groundwater resources have all taken their toll. The IPCC also believes that groundwater replenishment rates are probably falling overall as a result of climate change, though the effects are extremely varied and research is insufficient for any but local conclusions to be drawn. If this tentative assessment is correct, however, it would have implications for wetlands in those areas where replenishment rates are indeed falling.[15] Finally, where wetlands have been left undrained they have often been affected by pollution from agricultural or industrial sources or human settlements.[16]

The loss of wetlands would be tragic enough if it were only a matter of its impact on biodiversity and the beauty and variety of our planet's environments. However, this loss carries also serious implications for water quality.

As water runs off from higher ground, passage through a wetland area bordering a river or lake will subject it to a range of chemical, physical and biological processes which will alter its quality. Most of these alterations will be beneficial. They include the removal of nitrates and phosphorus, both of which are typically found in high concentrations in runoff from farmland. These present a direct danger to human and animal health and they can also result in eutrophication,[17] stimulating algae growth, which in turn reduces the water's oxygen content, killing fish and other aquatic life forms. In one study in North Carolina researchers found that wetlands reduced the nitrate content of runoff waters by 85 per cent. The same researchers estimated that the wetlands were even more efficient at removing phosphorus-bearing sediments.[18] It has also been shown that wetlands remove polluting plant nutrients and sediment by at least 30 per cent, in some cases entirely eliminating them.[19]

A broad survey of research into the value of riparian wetlands – those bordering rivers – conducted by scientists at the University of North Carolina in 1996 drew the following conclusion:

> Everyone who has studied riparian wetlands agrees that they provide many water-quality benefits. These are the areas that first receive and impede drainage runoff from developed lands. They are also the areas where shallow groundwater seeps into surface waters. Riparian vegetation traps sediment, removing harmful amounts of nitrate-nitrogen, phosphorus, and pesticides before they can enter streams. For all these reasons, there should be a strong effort to maintain or restore wet, vegetated buffers adjacent to streams.[20]

Wetlands are, however, threatened not only by development but directly by climate change. Only conscious human intervention can, under these circumstances, preserve the ones we have. Restoration will take a massive effort, but in the long term it may turn out to be the least costly way to maintain and enhance water quality.[21]

Increased Costs

The cautious global scientific consensus expressed by the IPCC is that '(w)ith respect to water supply, it is very likely that the costs of climate change will outweigh the benefits globally.' This will be partly because while both droughts and floods are likely to increase in frequency and intensity, 'infrastructure, usage patterns and institutions have developed in the context of current conditions'.[22] Demand for water may also increase, particularly from irrigated agricultural systems, as temperatures rise and precipitation becomes more variable, with patterns predictable in the recent past being disrupted. The 2006 HDR lists a number of examples to illustrate its conclusion that 'the overwhelming weight of evidence can be summarized in a simple formulation: many of the world's most water-stressed areas will get less water, and water flows will become less predictable and more subject to extreme events'. These include reductions in water availability in the Sahel and Southern Africa and East Africa, with production losses of 'up to 33% in maize and more than 20% for sorghum and 18% in millet' in the latter. As a report for the prestigious Stockholm International Water Institute (SIWI) warns, in an area 'already susceptible to food insecurity and where population will continue to grow, this is a drastic scenario'.[23] Globally food production systems will be disrupted, 'exposing an additional 75–125 million people to the threat of hunger'. Accelerated glacial melt will reduce water availability in East and South Asia and Latin America. South Asia will also suffer monsoon disruption, with more rain but much more concentrated in extreme precipitation events which will bring floods to parts of the region – or parts of the year – and drought in much of the rest. In its final example, the HDR warns of the effect of rising sea levels on river delta systems, where fresh water will turn brackish, transforming ecosystems and destroying livelihoods.[24] To summarise, climate change, joined with other factors including a rising population, will mean that the number of people living in areas suffering water stress globally will increase significantly.[25]

Of course, what is important, as we have tried to emphasise, is not merely the amount of water which comes out of the sky or down the river, but the amount which is actually available for use by agricultural, industrial and domestic consumers. This is also affected by infrastructure: by the size, design and positioning of storage and delivery systems. Nevertheless, if climate change means that less water is actually available as a result of a decrease in runoff or recharge of groundwater, the cost and difficulty of providing adequate supplies of safe water will increase. This will be equally true if the pattern of availability, the annual cycle of a river for example, is disrupted by climate change. Other aspects of change, such as altered water levels, may render existing infrastructure ineffective, leading again to higher costs. The IPCC lists the possible sources of additional costs linked to climate change's effects on water supply as follows:

- infrastructure renewal;
- the need to adapt to changing patterns of precipitation or snowmelt;
- the need to adapt to changes in rate of evapotranspiration;
- decreased water supply through irrigated systems coupled with increased water demand;
- the need for desalination to replace reduced supply of fresh water;
- adaptation to increased rate of flood and drought;
- the impact of alterations in levels of river discharge on hydropower generation. Although this is highly variable, with some regions and countries benefiting from stronger flow, hydropower potential for Europe as a whole is projected to decrease by between 7 per cent and 12 per cent by the 2070s, with Portugal, Spain and Bulgaria suffering a decline of between 20 and 50 per cent.[26]

The IPCC also noted that climate change would affect the demand-side of the water equation, increasing evapotranspiration and therefore plant water demand, which, coupled with more frequent drought, will increase demand for irrigation.[27]

This will all be expensive, with the most investment needed precisely in those areas least able to afford it, those which are already water-stressed and those where unsustainable agricultural systems are beginning to break down. Changes in peak flow periods for rivers will in many places increase the need for irrigation water. Hydroelectric systems such as those in China, India and the western United States are fuelled by melting winter ice. If the ice no longer forms, the water will not disappear completely, but its peak flow will change from spring and early summer to winter. In the past this may have been no bad thing, but the spread of air conditioning and refrigeration during the last half-century means that heaviest demand for electricity in warm parts of the world comes during the hottest months. Climate change will enhance this demand, just as it is reducing supply.[28]

Estimates of the cost of climate change must be taken with more than a pinch of salt. Running as they inevitably must into trillions of dollars, they become meaningless. All that can be said with any certainty is that it is going to be expensive. It is already expensive. And much of this expense is associated with water. Agriculture, energy and industry will all be faced with huge costs as they attempt to adapt to the warming and unpredictable climate. Politics will increasingly revolve around attempts by the private corporations which have profited from their climate-changing activities to shuffle this expense on to the rest of us.[29]

North and South

While conceding that warming may lead to higher food yields at cooler latitudes, the IPCC warns that yields of grain staples are likely to drop in tropical regions even at the lowest level of projected temperature increase and that anything above 2 degrees warming will have 'increasingly negative impacts in all regions'. As usual it will be the poorest who will suffer most, because '(r)egions where agriculture is currently a marginal enterprise' and which suffer under a 'combination of poor soils, water scarcity and rural poverty' are most at risk. The result is that 'even small changes in climate will increase the number of people at risk of hunger, with the impact in sub-Saharan Africa being particularly large'. The fact that the overall warming which is a central feature of climate change is accompanied by an increase in extreme weather events means that crop yields may drop even more, with heavy rainfall bringing about soil erosion or increased salination.[30]

In Europe, north and south will feel the effects of climate change quite differently, with the south likely to take the brunt, becoming warmer and drier, with all that means for agricultural production, hydropower, forestry and domestic consumers.[31] Not only will overall rainfall levels decrease markedly, but seasonal patterns will change: if and when the global temperature rise reaches 2 degrees, the long dry summer in Spain and the south of France may be extended by as much as five weeks, beginning a full three weeks earlier. Water shortages will simply become a permanent problem, especially in the densely populated but already arid areas of France's south coast, Italy and Spain. According to some projections, water shortages will become so acute that they will provoke mass migration northwards, with Europeans exercising their newly-acquired rights to settle in any EU member state.[32] There do not appear to be any EU plans to cope with this, at least none which has been made public. Yet water stress will also become more acute in central and eastern Europe as summer rainfall decreases and these are already the EU's poorest regions and the ones with the highest emigration levels. Implications for public order are clear, though apparently unrecognised, with the EU's authorities either considering that the problem is too remote, or that it is too explosive to be debated in public.

Even if crop yields initially increase as a result of warming, the north is certainly not immune to the negative effects of climate change. Biodiversity, ecosystems, soil, forests and recreational environments are all under threat as much as they are in the south, though the pressures are different. While the regions most prone to drought are all in the south, flood risk is spread much more evenly across the continent. Central Europe and the Mediterranean may experience a grim combination of decreased rainfall and what the IPCC report refers to as 'a substantial increase in the intensity of daily precipitation events'.[33] In simple terms: less rain, more storms. These changes will also have implications for water quality in Europe, with both extreme rainfall and drought increasing total microbial loads in fresh water and therefore the chance of outbreaks of water-borne disease in animals or people.[34]

Supply-side Measures

The question is, how well will Europeans, their governments and their Union be able to adapt to such changes? The IPCC points out that the usual response to increases in water stress has taken the form of 'supply-side measures' such as creating new reservoirs, but that this is increasingly discouraged in Europe by environmental considerations and high costs.

Waste water reuse is dogged not just, as is often supposed, by public squeamishness, but by genuine health problems and other difficulties. A rich city such as London can afford advanced water-treatment technologies to make waste water as clean and safe as any other water, but the processes involved are expensive.[35] Improvements in technology may well address the problem of expense and reuse of waste water clearly has a future. More affordable processes tend to produce water which is unsuitable for direct human consumption. Although the water remains usable for a range of purposes in agriculture and industry, supplies must generally be separated from those of clean water, creating additional expense. Urban areas can also use water which is unfit for direct human consumption: to irrigate parks and other public spaces; commercial uses such as vehicle or window washing; cooling systems and toilet flushing systems. Again, however, this creates a need for two separate water supplies. Apart from the health issue, reused water may create problems even when there is no possibility of it coming into direct contact with human beings. When used for cooling, for example, it may provoke scaling, corrosion, or, where nutrients or organic material are present, biological growth.

No internationally-recognised standards for water reuse exist, though the World Health Organisation has produced advisory guidelines and groups of neighbouring countries, for example around the Mediterranean, are making efforts to agree to them. The problem is enforcement. In the face of drought, respect for the law can be low. Unless countries competing with each other for the same markets for, say, high-value food crops can agree to draw up and enforce common regulations, waste water will continue to be used inappropriately.[36]

In the meantime, raw sewage is used to water an estimated 10 per cent of the world's irrigated crops. The sewage is rich in fertilising nitrates and phosphates and supplies are often abundant and reliable. The use of the sewage by farmers also has the advantage of less of it ending up in lakes, rivers and the sea. The disadvantages are that it carries the risk of disease for farmers and consumers and creates environmental problems due to the presence of toxic substances. Trying to ban the use of raw sewage is such an uphill task, however, that there are those who argue that a better approach would be to try to control it. Non-food crops such as cotton, or those which are likely to be heavily processed before being eaten, are more suitable for sewage-based irrigation than foods which may be eaten raw or lightly cooked. Fred Pearce, in his well-argued examination of the problems associated with water, *When the Rivers Run Dry*, cites a number of countries which, despite the fact that most of them are hardly wealthy, have developed forms of sewage recycling which remove pathogens and hugely reduce any risk to farmers and consumers. Juarez in Mexico, to take one of these examples, treats half of its sewage and uses the resultant effluent to irrigate 30,000 hectares.[37]

Desalination is energy-hungry and therefore both expensive and environmentally problematic, though it does have its advantages. Unaffected by drought, provided sea water is not polluted by hydrocarbons, the process will produce perfectly safe water. The traditional means of desalinating sea water is distillation, a simple technique known since ancient times. A technique known as reverse osmosis, in which sea water is forced through a filter, now tends to be preferred when new desalination projects are instituted, but 80 per cent of water desalination still relies on the older method. Technological advances, including reverse osmosis, have reduced desalination's energy costs, but these remain high, so the technique is most popular where energy is abundant and water scarce, such as the gulf states, or in a country like Spain where agriculture is potentially highly profitable. For every litre of drinkable water, a litre of brine is produced, which then has to be safely disposed of. Even if this were a straightforward mixture of salt and residual water it would have environmental consequences. In reality, however, chemicals used during desalination to reduce corrosion, the products of such corrosion as remains, and chemicals added in order to counter the build-up of scale within the desalination plants all find their way back into the sea. Desalination may well have a role to play, but it will never be a panacea. As things stand, only 1 per cent of the world's water supply and 3 per cent of its drinking water is produced by desalination. Given the large amounts of energy it demands, from the point of view of climate change it is part of the problem rather than part of the solution.[38]

Cloud-seeding is typical of the kind of high-technology solution the results of which are unpredictable and uncertain. As Pearce points out, what determines the weather is so complex that it is impossible to say if a particular attempt to provoke rain was or was not a success. There is evidence that it can actually backfire and stop rain from falling by causing the moisture to form too many droplets, which are consequently too small to fall as rain. And as cloud-seeding will not work unless moisture is already present, there is always the suspicion that what cloud-seeders are actually doing is stealing the rain that would have fallen on someone else's country or someone else's farm.[39]

'Virtual water' imports is the term used to describe the import of high water-use products by countries with insufficient supply to cover their needs. While this might work in a perfectly planned global economy with solidarity and equity as its prime values, in the real world in which we live it is a solution available only to the rich and exercised generally at the expense of the poor. For any but the very richest countries – and even, in the long term, for them – importing food which uses other people's water rather than tackling one's own country's water supply deficit is hardly a reliable way of ensuring food security.[40] An exception might be if more mountainous upstream countries in shared river basins were to trade the hydroelectric power they are able to generate for the products of the fertile land further downstream, but such high degrees of complex cross-border cooperation are rare.

There are undoubtedly technologies which are capable of improving the supply-side of the water equation without doing more harm than good. A number of techniques, some of them ancient, some new and innovative, are listed in separate examinations of the same issue by writers Fred Pearce and Mohamed Larbi Bouguerra: dew ponds in England, which gather moisture from the air and store it in clay-lined pools, dug by shepherds in the eighteenth and nineteenth centuries; the *zai* system in Burkina Faso, which exploits the natural behaviour of termites to capture and conserve water in deep pockets safe from rapid evaporation; greenhouses which condense water using sea water as a coolant; huge sheets of plastic mesh which harvest morning fogs in the Atacama desert, where it can go for years without raining (though these proved to be the victim of their own initial success, unable to supply sufficient water for the town whose growth they enabled; the result was that the government had to build a pipeline to supply it with water, and the plastic sheets fell into disuse and disrepair).[41] In China and India, the state encourages rainwater harvesting in cellars, cisterns, ponds and tanks, with enormous success in reducing demand on public supplies.[42]

Reducing Waste

Despite these successes, combining them with demand-side measures is essential. According to the IPCC such measures should include 'household, industrial and agricultural water conservation, reducing leaky municipal and irrigation water systems... water pricing... (and) introducing crops more suitable to a changing climate'.[43] The IPCC praises the European Union and its member states for incorporating 'watershed-level strategies to adapt to climate change' into 'plans for integrated water management'. The effectiveness of this and other aspects of the approach taken by European authorities, in particular where they are working within the guidelines laid down by the EU's Water Framework Directive, will be considered in Chapter 5.

Far more water is wasted than is actually used. Leaks, contamination, unnecessarily high rates of evaporation and pricing systems which do nothing to discourage overuse all play their role. 'Unaccounted-for water' is water lost from a delivery system before it reaches the consumer. It has been estimated that while this is as low as 15 per cent in North America and the UK, in developing countries it averages above 40 per cent.[44] In the OECD as a whole, the figure for municipal water loss is 30 per cent. The OECD believes that reducing such losses to the 10 per cent which it sees as 'the economically optimum level', could stabilise or even reduce water demand, enabling countries in some cases 'to avoid or postpone expensive infrastructure investments to expand water supply systems, including environmentally intrusive measures such as building dams and reservoirs'.[45] In the developed world, furthermore, clean drinking water is used for everything from washing the car to flushing the toilet, another form of waste. While, as we note above, laying parallel water delivery systems for different qualities of water would clearly be prohibitively expensive where water delivery infrastructure already exists, it ought to be possible to make this a planning requirement for new developments. Though this would require initially high levels of investment, savings over a long term would be immense and the benefits for drought-threatened areas incalculable.[46]

Reducing Demand from Agriculture

By far the biggest consumer of water is agriculture and it is to agriculture which we must turn if demand for water is to be reduced to the point of sustainability. The Green Revolution of the 1960s and 1970s brought higher-yielding varieties of major grain crops to countries which looked as if they would be unable to maintain sufficient production to feed rapidly growing populations. Unfortunately, these varieties also brought dependence on fertilisers and pesticides produced in the north, reinforcing neo-colonialism and restricting developing countries' policy choices. These crops were also thirsty, which meant a huge increase in irrigated agriculture and the consequent depletion of water reserves and reduction of the quantity of fresh water available on the planet as a whole. Deepening dependence on artificial fertilisers and pesticides, Green Revolution crops also had a major impact on the water supply in another way, with runoff pollution further reducing the amount which could be used.[47]

So the new crops enjoyed immediate success, but this came at the price of long-term unsustainability. Originally developed in the United States to boost the country's own agricultural production, they had already led there to much wider use of irrigation and the subsequent near-disappearance of once huge rivers in grain-producing states as well as the rapid depletion of the aquifers which feed them.[48]

Environmental activist George Monbiot points out that about half of the world's people live within 60 kilometres of a coast and that eight of the world's ten biggest cities are by the sea. As Monbiot explains, many coastal settlements

> rely on underground lenses of fresh water, effectively floating, within the porous rocks, on salt water which has soaked into the land from the sea. As the fresh water is sucked out, the salt water rises and can start to contaminate the aquifer... As the sea level rises as a result of climate change, salt pollution in coastal regions is likely to accelerate.[49]

The IPCC notes that although groundwater in shallow aquifers can be affected by variations in the climate, in most regions this has been far less of a factor during recent decades than has overexploitation.[50]

Reducing dependence on irrigation, using more efficient methods of irrigation, examining systems of crop rotation as well as which crops are grown where, developing agricultural and forestry methods which mimic nature's ecosystem inter-dependences and adopting practices which prioritise soil and water conservation: all of these things are possible and should be encouraged by policy-makers. Yet according to the UN's Food and Agriculture Organisation, the area of farmland under irrigation globally, having doubled during the twentieth century,[51] continues to increase and will carry on doing so. A large proportion of this expansion will take place in developing countries, where 75 per cent of existing irrigated farmland is found. Much of this will take place in water-stressed areas in South Asia, northern China and North Africa. Egypt is planning huge new irrigation projects, despite the fact that the Nile is already over-allocated to the extent that during dry periods of the year it no longer reaches the sea. In China, new irrigation schemes are being approved and funded even in the face of the obvious, visible fact that rivers are disappearing.[52] The IPCC suggests that increasing the productivity of irrigation water would mitigate the harmful effects of its spread and that such improvements 'are critical to insure (sic) availability of water for both food production and competing human and environmental need'.[53] It was estimated at the end of the 1980s that only 37 per cent of the water used in irrigation is absorbed by the crop plants targeted and this figure is unlikely to have improved.[54]

The IPCC lists what it calls 'autonomous strategies', ways in which farmers can respond to reduced water supply and contribute to conservation of the supply and its sources. These include the adoption of varieties or species more tolerant of heat shock and drought; modification of irrigation techniques; adoption of water-efficient technologies to harvest water, conserve soil moisture and combat contamination through silt or salt; improved water management to prevent waterlogging, erosion and leaching; and modification of crop calendars.[55] It goes on to advise governments and those devising development programmes to promote such adaptations.[56]

A report prepared for SIWI concentrates its attention on 'water harvesting' to improve supply for rain-fed agriculture which, as it points out, even in water-stressed areas still constitutes the majority of farming activity. Globally, it amounts to 80 per cent of all agriculture, and though this is currently falling, the potential for it to fall much further simply is not there. There is a severe limit to how much water can be withdrawn through irrigation and in many areas this has already been reached or breached. If we are to find a way out of the water crisis, therefore, we must find ways of using rainfall more efficiently.[57]

SIWI's researchers found that in eastern and southern Africa, land degradation and the fact that the rain, when it falls, often does so as a 'high intensity rainfall event' meaning that farmers do not get the full benefit of the rain that does fall. The same thing happens anywhere in the world if you do not hoe your garden: rain falls and runs off the surface, or at least does not reach the plant's roots. The amount of water actually used by plant crops where soil is degraded, as in much of the two regions cited, is generally as low as 15 to 30 per cent, and in some cases only 5 per cent. Even water reaching the 'root zone' may not be fully utilised due to poor soil, because to absorb water the plant needs nutrients to be available in the soil.[58]

Improving effective water supply is therefore clearly possible and may have a knock-on effect. The researchers noted that farmers' reluctance to invest scarce resources in fertilisers, improved varieties of crop plant or pest management stemmed in part from the fact that they experienced yield reductions on average almost every other year. Recent research has shown, however, that if enhanced and more reliable water supply is available, coupling this with actions to improve soil fertility, manage pests and backup services to farmers – including ones which improve their own skills – can lead to a doubling or more of yields.[59]

In considering each of the above approaches, however, it is important to realise that there is no single perfect solution and a few which should never be applied at any time, anywhere. Numerous variables – to do with economic and cultural factors as well as those which are directly meteorological or hydrological – will affect the answer to the questions posed by climate change, including water scarcity. As Constance Elizabeth Hunt notes, what we are dealing with is perhaps better seen as 'a series of water cycle disruptions operating at a variety of (often interacting) scales' rather than a global problem which must therefore have a global solution. In this way, 'the imposition of inappropriate technologies on cultures and ecosystems' can be avoided.[60]

Tackling Climate Change

As well as considering what can be done about meeting the demand for water in a world where the climate is changing rapidly, however, we should first look at what, if anything, can be done to address the problem of warming itself.

There are just two schools of thought about this. Firstly, there is the conservative technocratic school, which wishes to carry on business much as usual by finding new sources of energy which will prove less damaging to the planet. In this way, we in the west can continue having lots of toys, living in places nature intended only for snakes and spiders, and financing frequent military adventures in faraway places. Meanwhile, the people in the faraway places who learn to do as we tell them can also have these things, and global equality will eventually be reached in the form of one household, one multimedia en suite entertainment system. Unfortunately for those people who believe we are living in a consumerist utopia and would like to see this happy valley spread, their solution to the problem of climate change simply won't work.

The second school of thought is the one to which the authors belong, the one which says that only radical changes in the way that we see the world and relate to it will bring about the sort of global transformation necessary if we are to be able to continue to live in a more-or-less civilised way. This will not be a matter of 'making sacrifices', but of rediscovering the prime importance of human dignity, equality and solidarity and applying these values to the way we organise our economies.

We cannot continue business as usual because the technological, technocratic solutions proposed simply will not work. And in many cases an understanding of the hydrological cycle that underpins all known life is the key to an understanding of why these neat, essentially conservative solutions will not work.

Let us begin with the obvious: hydropower. We have already seen that the dams required to generate electricity in this way in almost all cases come at the price of social and environmental disruption more or less in ratio with their size. The defence of this is that it is a necessary trade-off. Hydropower is clean and does not warm the planet and is therefore a fine substitute for carbon fuels. Shame about the drowned villages, the floods, the disrupted agricultural systems, the massive environmental damage and the reduced biodiversity, but these are prices we have no choice but to pay. Unfortunately, or fortunately, depending on where you're standing, all of this turns out to be wrong. In fact, hydropower dams contribute to global warming.

The forests, wetlands and soil which the hydropower dams drown both consume and emit greenhouse gases: carbon dioxide and methane. The balance depends on the type of ecosystem: in mature forests and grasslands consumption and emission are likely to be equal, but in some systems soils consume more methane than they release. Flood a forest and you lose a major carbon sink. Decomposition of flooded plants and soil releases carbon which will eventually find its way into the atmosphere. Tropical wetlands are a source of methane, but flooding them permanently increases emissions of this greenhouse gas. Anti-dam campaigner Patrick McCully explains that gases from reservoirs 'can be emitted through continuous diffusion into the atmosphere from the surface of the water; in sudden pulses when deep water in the reservoir rises to the surface in cold weather... and from deep water being discharged through turbines'. This last is because 'warm, nutrient-rich and severely oxygen-depleted water at the bottom of.... shallow reservoirs creates ideal conditions for the methane-producing bacteria which feed on decaying vegetation'. The slow rate of decay means that 'methane emissions are fairly constant over time and do not decline significantly as the reservoir matures'. Compared to carbon fuels, results are mixed. A study of two projects in Brazil showed that while one, Tucurui, was more carbon-friendly than a coal-fired plant generating the same amount of electricity, emitting only 60 per cent of what the latter was responsible for, it emitted half as much carbon again as a gas-fired power station. The other, however, Balbina, 'had *26 times more* impact on global warming than the emissions from an equivalent coal-fired station'.[61]

Nor is this the whole picture. To the greenhouse gases emitted from the reservoirs must be added emissions resulting from the dam's construction and the production of the materials from which it is constructed, as well as those resulting from 'changes which the dam encourages, such as deforestation, the conversion of floodplain wetlands to intensive agriculture, the adoption of irrigation on once rainfed lands, and the increased use of fossil-fuel-based artificial fertilizers'.[62]

Unlike hydropower, which is vulnerable to climate change and, as we have seen, offers no reliable net reduction of greenhouse gas emissions, nuclear energy generation does offer some kind of contribution to a solution to the problem of global warming. The nuclear industry, which exists only through massive public subsidy and must therefore maintain a vigorous propaganda campaign in the face of widespread scepticism, unfortunately appears to have scored something of a hit with its association of nuclear power generation with 'clean' energy. If this were not so dangerous, it would be hilariously funny, and one wishes that Jonathan Swift were still around to do it justice.

The problems with nuclear energy are well-known and do not need our detailed attention. They may be listed as follows:

- the impossibility of finding or developing a safe method of storing the resultant waste, which consists of poisons which will continue to be deadly for tens of thousands of years;
- the enormous costs involved in nuclear power generation, a large slice of which is associated with decommissioning power stations when they come to the end of their useful life;
- the likelihood, or indeed inevitability, of deadly accident, military strike or terrorist attack;

- the likelihood of theft of materials leading to the production of the so-called 'dirty bomb';
- the resultant additional costs due to the need for effective security measures and in some cases the encroachment on civil liberties which these entail;
- the fact that if all of this money were spent on genuinely renewable alternative energy sources and effective conservation measures, a very large contribution to solving the problem of providing the energy we need without destroying the planet really would be made.[63]

Bio-fuels may turn out to have been an even worse idea than nuclear power. With oil prices rising precipitously from 2003 onwards, the bio-fuel revolution that had been bubbling under for some years took off. For many years before that, governments had been supporting bio-fuel production with various forms of financial aid and with the twin attractions of high fuel prices and state support bio-fuel production could hardly fail. Within two years of oil prices beginning to experience rapid inflation, ethanol production had risen worldwide by 13 per cent, the US's by 20 per cent and Germany's by 60 per cent. By the beginning of 2006, around 6.7 billion litres of new fuel-ethanol capacity was under construction in the United States, more than twice the figure of January 2005. Biodiesel was no different, with total production capacity within the EU almost doubling and explosive growth in Australia, Canada and the United States.[64]

Though market forces have played a role in making bio-fuels attractive, it is government policies which have done most to promote their spread. As a study of bio-fuels subsidies by the International Institute for Sustainable Development (IISD) remarks, the wide range of alleged benefits has meant that they have attracted an extraordinarily wide and varied range of support. The result has been the creation in the EU and Switzerland, North America and Australia of a friendly fiscal and regulatory environment. The benefits which supporters have attributed to bio-fuels include boosting agricultural production and employment and thus rural development, reducing greenhouse gas emissions, improving urban air quality, enhancing energy security, more favourable trade balances and providing economic opportunities for developing countries. As the report comments, it is not that there is no possible relationship between bio-fuels and each individual item on this list, but that they are in some cases mutually contradictory, that 'not all of these objectives can necessarily be pursued at the same time through policies supporting a pair of fuels'. The report adds, with splendidly euphemistic understatement, that '(t)he political economy of public transfers is such that the risk of public policy being coopted in support of private ends is and will remain great'.[65]

The report concludes that bio-fuels have not met their objectives and that the very least that is needed is a 'top-to-bottom' rethink of the overall rationale for supporting bio-fuels. There should, it argues, be no new subsidies and the ones which exist should be phased out. Mandated use of bio-fuels is, it concludes, driven by pressure from the industries which stand to benefit directly. The report ends with a plea for a moratorium and further research into the consequences of growing plants for fuel.[66]

Yet at the end of 2007 the United States Congress passed the Energy Independence and Security Act requiring the nation to produce almost 57 billion litres (15 billion US gallons) of corn ethanol annually by 2015, 10 per cent of the country's needs. This level of production, one critic of the Act has written, 'would require massive, permanent increases in the amount of land sown to corn, as well as ramped-up water consumption and pollution'.[67] The European Union has similar policies, with a target of 5.7 per cent market share by 2010 and 10 per cent by 2020, though a recent apparent outbreak of cold feet may see these diluted.[68]

Rising global food prices, due to no small extent to farmland being turned over to bio-fuel crops, have provoked a re-examination of what has turned out to be, to look kindly on the motives of those involved, at best a serious error of judgement.[69] The fact that bio-fuels are the major reason for rapid and destabilising food price inflation is not the view of sceptical green politicians or activists but that of both the IMF and the World Bank, one of whose leading economists is on record as saying that 'biofuels and related low grain inventories, speculative activity, and food export bans pushed prices up by 70 percent to 75 percent'.[70]

Far from making any positive contribution to combating climate change, the expansion of bio-fuels has resulted in increased pressure on land and water resources. Deforestation to make way for the colossal levels of production which would be needed in order to make any significant dent in atmospheric carbon would more than offset any gains.

Carbon sequestration offers an alternative approach to that proposed by advocates of hydropower, nuclear power and bio-fuels, or one which would complement them. Carbon dioxide would be removed from the atmosphere and stored in various places known as 'sinks'. Sinks may include oceans, deep underground strata or cavities, perhaps even former aquifers which have been over-exploited until empty. None of these proposals has been sufficiently tested and computer models give mixed results, with the potential for disaster, as with stored nuclear waste, obvious and difficult to argue away.

The alternative would be to plant more trees or other vegetation, which would, if handled correctly, bring side benefits in the form of reduced soil erosion, enhanced biodiversity, useful forest products and even recreational facilities. If handled carelessly, however, it would not only fail to deliver such benefits but would do little or nothing to reduce carbon emissions over any realistic term. The only way to ensure that increasing the Earth's vegetative cover would be done effectively would be to remove the profit motive from it completely. Instead of this we see attempts to make carbon sequestration through tree planting a tradable commodity. What the plans have in common with the untried sinks in the sea idea is that their unintended effects will be unpredictable, unprepared for and probably extremely dangerous. If we have learnt anything over the last two centuries it should surely be that you mess with natural cycles at your peril, not because there is something 'sacred' about 'nature', but because you are dealing with highly complex systems about which much remains poorly understood.[71]

Carbon sequestration is, however, achievable by methods which involve much less risk of disruption as well as not leaving our descendants with an unmanageably dangerous legacy. Atmospheric carbon levels can be reduced through the halting – or more realistically, the slowing – of deforestation, the promotion of natural forest regeneration and the promotion of sustainable reforestation schemes. In addition, changing to no-till or conservation tillage systems can counter soil erosion while reducing emissions of carbon from the soil into the atmosphere. Reducing the number of cattle in the world would also help and such a reduction would undoubtedly follow the reduction of subsidies to make meat and dairy products reflect the true costs of their production, which would have incidental benefits for the health of consumers in developed countries.[72]

Climate change must be addressed not through technologies which seek to maintain the status quo, but by rethinking the way in which we provide ourselves with the lighting, heating, cooling, freezing, transport and other essentials either of life itself or of a decent and rewarding existence. First and foremost energy could be much more efficiently used and the good news is that there have already been major improvements in energy efficiency in most parts of the world in the last four decades. The same UN study which estimated that nearly two-thirds of primary energy is lost before it can be used, estimated that the first two decades of our century 'will likely see energy efficiency gains of 25–35 percent in most industrialised countries and more than 40 percent in transition economies'. These would be achieved by 'dematerialization' – the reduction in the quantity of material used to produce a unit of economic output – as well as recycling, making 'energy efficiency ... one of the main technological drivers of sustainable development world-wide'.[73]

During the early years of the twenty-first century, rapid development has been seen in energy-efficient technologies, more efficient generating techniques such as combined heat and power co-generation and the use of what was previously waste heat in industrial processes.[74] Encouragement of such developments not only through the provision of information, but through the structuring of tax systems would be of much more assistance in the fight against global warming than would the construction of yet another megadam.

Finally, renewable energy sources such as wind, solar power, tidal power and geothermal power are capable of making an enormously greater contribution to the satisfaction of energy demand. Fuel cells which combine hydrogen with oxygen and make use of the energy released in the process will enhance the conservation effect. Whatever technical and economic difficulties may stand in the way, the only real solution to the problem of global warming and the accompanying manifestations of climate change will involve greater energy efficiency, energy conservation and the fullest possible development of renewable energy sources, enabling the minimisation and eventual phasing out of carbon-based sources. Limiting climate change in this way has the potential to make the job of managing the supply of water to a growing global population feasible.[75]

5

THE EUROPEAN UNION WITHIN ITS BORDERS: WHY PRIVATISATION? THE IDEOLOGY BEHIND THE THEFT OF PUBLIC PROPERTY

'They hang the man and flog the woman
Who steals the goose from off the Common;
But let the greater criminal loose
Who steals the Common from the goose.'[1]

Political theorists of the left have for some time been aware of a process that they have termed 'new constitutionalism' or, slightly less abstrusely, 'depoliticisation'. The argument of thinkers such as Stephen Gill and Andreas Bieler, leading lights of what is known as 'neo-Gramscianism', is that neo-liberalism works in part by removing vital economic decisions from the realm of politics and thereby from any democratic or popular input. Legal and technocratic mechanisms replace elected institutions or state employees answerable to them. Elections themselves are reduced to beauty contests, or, at best, to consultations affecting issues which do not touch the core of the economy. These issues may be of importance. In fact, the more important they are the better they will disguise the erosion of democracy in the economic sphere. They may therefore include such vital matters as abortion rights, religious education in schools, toleration or otherwise of homosexuality, definitions of and responses to crime and whether foreign languages are taught in schools. They may not, however, include decisions on public versus private ownership, market access or state subsidies to industry or agriculture. These are no longer defined as political questions.

This profound transformation of politics has occurred within a generation. When the older of the two authors of this book first voted, in 1974 at the age of 19 in a working class suburb of Manchester, England, the choice before him was clear enough. While no-one was offering either a socialist or a 'free market' utopia, most people could have told you that the Conservatives stood for a greater element of private ownership within Britain's 'mixed economy', while Labour stood for more social or public ownership. The political system was scarcely viewed less cynically than it is now, and people understood well enough that many barriers stood between the expression of the popular will and what actually emerged from government, but few would have argued that the people did not have the right to make this choice. Both nationalisation and denationalisation had their passionate advocates. No doubt each thought the other group misguided or self-serving, but their right to hold these views (and should they prevail at the ballot box have them implemented) was not in question. If the Conservatives had proposed that water supply be given over wholly to private corporations, for example, the choice of whether to let them put this into practice would have been regarded as being legitimately that of the British people and no-one else.

By the end of the century all of that had changed. Economic policy is increasingly insulated from the democratic process. The tools with which this insulation has been achieved include the World Bank, the International Monetary Fund and the World Trade Organisation. Regionally, however, the most powerful instrument of depoliticisation is the European Union. Each of these institutions began as something quite different from its present manifestation, though in the case of the EU the seeds of the removal of economic decision-making from the contested space of politics are evident even in the Treaty of Rome itself, the 1957 instrument which founded the European Economic Community. This insulation of economic policy is the key feature of neo-liberalism. It has a number of ideological pillars.

Firstly, it is based on an assumption that private ownership in a competitive market is inherently more efficient than public ownership. The fact that, as we shall see, there is no evidence whatsoever for this is of little importance, as the same people who own the means of production and distribution of goods also own the vast bulk of the means of production and distribution of information and 'knowledge'. This has enabled them to create the widespread impression that the inherently greater efficiency of private competitive markets is an established fact.

Secondly, it relies on a redefinition of the word 'freedom'. The 'free market' becomes, under this redefinition, an essential element of a 'free society'. The freedoms to trade, to produce, to consume and to act within the economy without restraint thus become equivalent to such traditionally recognised freedoms as the freedom from arbitrary arrest, freedom of speech and freedom of assembly. This is significant, because it means that they need no other justification, but are an end in themselves. We can understand this if we compare it to the freedom to gather peaceably. Whilst it is possible to give goal-oriented arguments for freedom of assembly, to argue that only by allowing and encouraging the free expression and exchange of ideas can a society maximise its intellectual potential, few would feel the need to do so. It is simply right that if people wish to assemble peacefully they should be allowed to, whether their intention is to play football, have an *a cappella* sing song or discuss the failings of the government and what can be done about them. It is important to understand that neo-liberal ideology rests on a redefinition of economic 'freedom', one which makes of it a freedom absolutely comparable to freedom of assembly in these respects. Both freedoms may conceivably be abrogated, but only under the most extreme circumstances, such as in times of war, and in such cases it is up to those doing the abrogating to justify their actions. As Colin Green, Professor of Water Economics at Middlesex University in London says, the drive to privatise water in the UK in the 1980s 'derived from the then dominant ideology within economics which had a strongly libertarian focus', seeing 'privatisation as a desirable end in itself rather than as a potential means to some higher ends'.[2]

Thirdly, in neo-liberal ideology, corporations are the potential suppliers of all of our wants. State-owned industries are inherently inefficient and small-scale suppliers can never enjoy the advantages – economies of scale, conglomerations of expertise, global reach – which enable corporations to deliver the highest possible levels of service or goods at the lowest possible price. The role of government in the economy should be restricted to that of referee, a body which is invested with the power to deal with crimes against freedom (which is to say, monopoly practices and other unfair restraints of trade), crimes which may be carried out by corporations themselves, by trade unions or by other entities. The problem is that the state, the obvious choice of umpire, is itself one of the worst culprits. Because of this, a supernational body which will not be tempted to cheat either to its own perceived advantage or that of its own national corporations is the ideal instrument. The role of small companies should be restricted to sub-contracting, or dealing with tiny niche markets to which large corporations may be unresponsive, though these will be taken under the corporate wing once they become established.

Fourthly, in an extension of what Milton Friedman, one of the founding fathers of this ideology, called 'market-preserving federalism', freedom of movement of labour, capital, goods and services must be as absolute as practically possible. The purpose of this is to undermine the power of any state to place conditions on these factors of production within its jurisdiction. A level of corporate taxation which is unacceptable to a corporation will result in that corporation delocalising to a state which imposes more reasonable levels. High wages will attract labour from elsewhere, automatically lowering the level of wages, for any 'artificial' minimum wage will again cause corporations to move house. Any restriction on the movement of goods or services will reduce the efficiency of the economy by preserving the activities of companies, or even of whole sectors, which can be more efficiently carried out elsewhere.[3]

The 'four freedoms', as these are known, thus serve to discipline states. That electorates may have voted to modify one or more of these freedoms is of no interest to those who hold this ideology. This may be a position with which we can disagree, but it is not illogical. We surely all have a belief in fundamental freedoms or rights which are beyond the scope of electoral politics. We would hardly allow that slavery could be reintroduced by popular majority vote, for example. This is one reason why most countries have constitutions which cannot be easily amended. The neo-liberals simply want to add the four freedoms to this group of concerns, to matters beyond politics. A government's rights do not, in this view, extend to placing restraints on the movement of capital, except to protect against such things as fraud, no more than they extend to persecuting ethnic minorities within their boundaries. Instead, a government's job is to provide a sound environment in which corporations will wish to do business and investors will feel comfortable placing their money. By making capital mobility absolute throughout its territory, the European Union is thus expressing its neo-liberal ideology. By making it a condition of membership and part of the conditionality it imposes on loans, the IMF does the same. By seeking to guarantee the mobility of transnational direct investment, the WTO is attempting to do the same globally. The EU also plays this game outside its borders, as we shall see in Chapter 7.

In the past, many conservatives accepted that certain economic sectors must remain within the remit of the state. This was particularly the case after the Second World War, when huge support for Communist Parties in Italy and France, and the emergence of strong radical wings in social democratic and labour parties elsewhere in western Europe, forced the owners of capital to come to terms with the aspirations of working people. In and among the countries which would become and which remain the most developed member states of the European Union, a political consensus emerged which included support for, or at least acceptance of, state ownership of essential services. Water being the most essential of all, it was taken into public ownership in almost every country of this region where this was not already the model. Those parts of the service which remained outside public ownership were subjected to strong supervision and control, not least when it came to permissible rates of profit.

This thinking, however, is anathema to neo-liberals. This again is logical. Their belief in the market revolves around the price mechanism and to be consistent it must be absolute. Water must be treated as a commodity like any other, because the price mechanism is always and in all circumstances the best way to govern the distribution of resources. If water is not best governed by the price mechanism, then this would imply that the price mechanism is flawed, which cannot be.

The idea of the market and the price mechanism being the best instruments for the distribution of water has other problems, however. The market works best in conditions of intense competition. Provided some authority succeeds in eliminating monopoly practices, price-fixing and so on, competition can serve to guarantee that the consumer pays a fair price and that the most efficient producers and distributors prevail. Objections to the market as arbiter of all things are manifold, despite this, but let us take neo-liberalism for a moment on its own terms. The crucial phrase here is 'in conditions of intense competition'. Without such competition, goods must be subject to some form of price control, unless they can be wholly done without.

Water cannot be done without at all, even for a few days. Very few people have any serious alternative to depending on the public supply, if they are lucky enough to have one, and almost everyone in Europe is in that happy position. Their tap, or at worst their village or neighbourhood standpipe are always going to be their major, if not their only, source of water. So the first problem is simply this: that there are no alternative products.

The second problem is that without the introduction of the most artificial devices, devices which mimic competition rather than really providing it, water is a natural monopoly. The market simply does not function when it comes to an essential product for which there can really only be one supplier per region or town. The French thought that they had an approach which solved this problem, but we will see below what has become of that idea.

How Ideology Becomes Practice:
The EU Legal Framework and the Drive to Privatisation

The European Commission has repeatedly claimed that it is neutral on the question of public versus private ownership. It can be demonstrated, however, that this claim is disingenuous, that in reality the Commission is a partisan enthusiast for ever-increasing privatisation, including essential services. Sometimes, this is quite explicit. More often, however, it is woven into the fabric of EU law. This can be seen most clearly in the Treaty on European Union's approach to competition policy and the internal market. David Hall, of the Public Service International Research Unit (PSIRU), which has conducted a long-standing series of studies of EU policies as they affect the vital services, explains that this affected neutrality is made necessary by an 'obscure clause in the (EU) Treaty which says in effect that member states can decide what they like on the question of ownership'. This 'does in practice prevent them from being straightforward advocates of privatisation. It's correct to say this claim is disingenuous because they do in practice do a lot of things which favour privatisation. PPPs and competition policy and the internal market are all good examples of that.'[4]

While the EU is prevented from actually banning public ownership, this has not prevented it from eliminating many of its advantages. A major potential benefit of public ownership of an essential service, for example, is the ability to cross-subsidise. In postal services, rail and other public services, and in relation to energy and water, income from profitable sectors can be moved around to cover unavoidable losses elsewhere. Decisions are taken on the basis of the general or social good and in western Europe at least they have long been taken by democratically elected governments. Rural railway lines are run at a loss, subsidised by profits from popular inter-city services. A universal, one-cost postal service is maintained, so that a letter may be posted from Shoreditch to Shetland as cheaply as it can to another part of London. Poor households have their water bills subsidised out of higher rates charged to those who can afford it. To some, this is the basis of a civilised society in which solidarity enables, for example, all children to have a reasonable start in life.

To others it's anathema, an attitude which is difficult to understand. The criticism is that such a system may hinder transparency. As we shall see in the next chapter, however, the Water Framework Directive does indeed abolish such cross-subsidy systems.

David Hall of the PSIRU points to the electricity sector to illustrate the potential effects of this:

> In the electricity companies there's quite a lot of public ownership around. It's permitted as long as they behave like privately owned companies. What you can't do is create vertically integrated monopolies (in which) you can build in all sorts of cross-subsidies. This is impossible under liberalisation.

Hall points out that in a liberalised system, a customer who is charged more for a service in order to subsidise another user can simply take their business elsewhere. This is, he says 'one of the worst aspects of the Water Framework Directive, which says that there can't be any cross-subsidy between broad categories of user – agricultural, business, household. It makes it technically illegal to charge say businesses more so that you can charge households less.'[5]

What Hall refers to as 'this erosion of the right to establish cross-subsidy' has been enacted in all of the liberalised sectors. Water, however, is not fully liberalised. Yet despite the fact that the EU has stopped short of enforcing compulsory competition in the sector, they have, he says, moved to put an end to cross-subsidy. Linked to this is the fact that although the WFD does not require full cost recovery, it requires 'movement towards full cost recovery'. The difference between these two positions seems somewhat theological and sure enough the Commission is acting 'as if it did say that it requires full cost recovery'. Hall gives the example of Hungary, which has been told that it cannot continue to invest in water infrastructure from general taxation. 'You've got to shift to financing it through water charges.' He confirms that Ireland, which finances its water supply entirely from general taxation, has been told the same.[6] This sets up a potential conflict between different elements in the Treaty, which makes it clear that taxation lies within the competence of the member states and not of the Union or its institutions.

PSIRU produced an analysis in 2003 of the approach to the water industry taken by the European Commission Directorate General for Competition. Known as DG-Competition for short, this branch of the EU's executive had recently produced a report which provides a perfect illustration of everything we have said above about the ideology behind privatisation, liberalisation and deregulation. *Study on the Application of Competition Rules to the Water Sector in the European Community*,[7] was, appropriately enough, outsourced, in this case to a UK-based privately-owned research organisation, WRc, which does most of its work for the UK's Environment Agency.

The report indicated that the Commission's interest in the water industry was stepping up a gear. Based on the usual assumption about the relative efficiency of the private sector, it signalled what PSIRU rightly feared would be 'a continuing agenda from DG Competition to transform the sector', the emphasis being on providing opportunities for corporate business, rather than the interests of consumers, workers or the environment.[8]

Although the WRc report states, as is commonly claimed by the Commission itself, that the 'process of looking to see how EC competition rules could be best applied in the water sector is about making a contribution to promote greater efficiency and higher levels of service in the sector' and 'not about the promotion of private sector participation', its entire approach is based on an assumption that such participation is essential to 'efficiency'. Yet as PSIRU points out, 'empirical evidence on the relative efficiencies of public and private operation in water – as in other sectors – is, however, consistently neutral'.[9]

The approach favoured by DG-Competition makes no distinction between private and public monopoly. Yet there is indeed a clear difference. Whereas a private sector operator is in business to make profits for its shareholders, a public sector monopoly may have a range of motivations. Pricing policies of publicly owned enterprises may be revenue-neutral, for example, in order to guarantee the greatest possible social equity. Or they may be required by law to invest any surplus in improving infrastructure, working conditions, or quality of service, or in environmentally beneficial ways. This can be achieved under a private-ownership model only by means of public subsidies. There is no getting round the fact that whereas, ideally, private sector water service suppliers can be required to meet obligations just as exacting as those imposed on publicly owned suppliers, the problem is that they have no interest in doing so, other than to avoid prosecution. And if they are successfully pushed into delivering services of the same quality as a comparable publicly-owned enterprise, they will inevitably be more expensive, obliged as they will be to add a percentage on top in order to reward their shareholders. As PSIRU notes, the WRc report 'treats public sector operators as though they can only be profit-maximising commercial entities with an unusual shareholding structure'. Yet in reality, 'public ownership is a mechanism for seeking to ensure that public interest considerations replace monopoly profiteering'.[10]

Public ownership of services should ideally form part of a socio-economic policy which enables vital services and products to be delivered to those who need them at a price which can be afforded – as this was once rather famously expressed: 'From each according to his ability, to each according to his needs!'[11] Public ownership also enables the sector in question to form part of an overall economic strategy designed to maximise social benefits.

To give an example: a governmental authority may have a number of policy priorities relating directly or indirectly to water. These may include ensuring that everyone receives a supply at a price he or she can afford; ensuring that the 'product' is of the highest possible quality and, within the EU, that it meets the sometimes exacting standards rightly imposed by the various European directives; that it is not wasted; that reasonable priorities are imposed on its supply – hospitals before golf courses, for example, or a basic minimum for all before you get to fill your garden swimming pool; that this supply is organised in ways which are harmonised with other aspects of policy, such as environmental protection and broader public health concerns, and that systems are in place for addressing emergencies such as droughts and floods. The use of water for purposes other than consumption – for example to produce power – must also be taken into account.

These different policy priorities may sometimes conflict. Ensuring that water is affordable and ensuring that price structures encourage its conservation do not fit comfortably together, for example. If they are to be made compatible it must be on the basis of a broad social debate and through popular input by means of democratic structures and in the light of informed advice from experts. Everyone involved, though their immediate interests may clash, should be working towards a single goal which might be summarised as an effective, fair, efficient, environmentally-sound system of water delivery and sanitation. Introducing an element into this whose only rational attitude must be, 'that's all very well, but I must have my 10 per cent' seems an odd way of going about the pursuit of such delicately-balanced goals.

Pricing within a publicly owned industry may also be made to serve broader goals than those immediately relevant to the sector in question. Surpluses – 'profits' – from water might be used for purposes which have nothing immediately to do with water *per se*, provided the water supply and sanitation systems are adequately funded. 'Externalities' – the social and environmental costs incumbent upon the organisation of an effective system of water supply and sanitation – can be included in water bills or paid for out of general taxation, or through some other system. Again, what is vital about such decisions is that they are taken democratically and by bodies structured to encourage popular participation. Do we want a new water-based leisure park? lower water bills? restoration of natural flood defences? Such matters must be decided democratically and not by corporate directors.

Competition as such is, in economic terms, neither good nor bad: the decision as to whether a sector would benefit from it should be a political decision freely taken by democratic and popular institutions. In the case of a natural monopoly such as water, however, competition seems to be the most roundabout way of improving efficiency, one unlikely to succeed and extremely difficult to ensure. This is demonstrated by the fact that water supply and sanitation internationally exist almost entirely outside the competitive economy. Two companies share around 70 per cent of the global private sector market. The behaviour of these companies is persistently anti-competitive. They:

> act together in joint ventures with each other and their competitors in a number of countries...prefer to keep contracts secret...have had executives convicted of corruption in France, Italy and the USA...and defend their existing contracts with considerable legal energy.[12]

Those who favour privatisation will often argue that while competition *in* the market may be impossible, it can be replaced by competition *for* the market. David Hall disagrees, pointing to what happens in France, where this has long been the system, though compulsory competitive tendering is a relatively recent refinement:

> The weakness of this argument is that firstly it's competition at just one moment for an enormously valuable twenty-five years and therefore it's competition which is very limited in time and very vulnerable to constant renegotiation and to becoming subject to political pressures. In practice the amount of competition is simply very low.

According to Hall, experience has shown that 'when you compete for the market your incentive is to devise tactics that result in your being awarded the contract'. These tactics 'may actually have nothing to do with the level of service'. The obvious assumption to make, for anyone not in the know, would be that the basis for competition for the market is 'all to do with who is offering the best value of service in terms of quality and price over the next 25 years'. In practice however 'what will happen is the framing of bids and of the evaluation package, and competition will be around these'. Costs are invariably underestimated. Corruption *per se* is thus only one means of channelling taxpayers' money into private pockets in a way which most definitely does not offer value for that money.[13]

Around the same time that DG-Competition was commissioning a report clearly designed to justify putting pressure on member state governments to privatise their water supply and sanitation services, the Commission's Directorate-General for Internal Market and Services – known as DG Markt (*sic*) – was casting its eye on the same industry. Its general paper on internal market strategy for 2003–2006 again demonstrates the way in which neo-liberal ideology has come, in Brussels, to be seen as equivalent to the truth.[14] It is a truth, as we have pointed out above, that no longer needs empirical justification, it is a philosophical principle rather than an evidence-based position.

It should be noted, moreover, that as little water is actually traded between countries, DG Markt should have little to say on the subject. To understand the role of internal market policy, we can compare water to that famous mythical good, the widget. DG Markt's job is to ensure that nobody discriminates in favour of their own national widget industry. This is supposed to ensure that the most efficient widget companies in Europe get the most contracts to supply them, which leads to 'savings'. Also, a single internal market should, so the theory goes, ensure that the widget sector reaches the appropriate degree of concentration, enabling further savings through economies of scale. DG Markt is always concerned to avoid 'fragmentation' of industries, which is different from healthy competition and can only occur through market 'distortions'. For example, if major purchasers of widgets – such as local authorities or even national governments – favour their 'own' widget industries, inefficient producers will remain in business. If they move to prevent takeovers 'in the national interest' – if for example British Widgets plc is protected from takeover by its super-efficient rival Algemene Dingetjes Nederland N.V. – the result will be 'fragmentation' and a reduction in the overall efficiency of European widget manufacture.

There are counter-arguments to all of this, but we need not concern ourselves with them now, as none of it can be applied to water. All that can be achieved is the replacement of one kind of monopoly with another. According to the Commission, 'action may be required' because 'the water sector...remains fragmented'.[15]

Yet given that it is a natural local monopoly, how can it possibly not be 'fragmented'? As PSIRU points out, 'water is run by munici- palities in all EU countries', so that a more positive description than 'fragmented' would be possible: for example 'community- based'.[16] DG Markt, while claiming to be 'neutral' on 'the question of ownership of water and water services', goes on to say that water is a sector 'where there are potential gains to be had from modernisation'.[17] As this cannot possibly mean modernisation of infrastructure, which has nothing to do with DG Markt's area of responsibility, it can only be, as PSIRU notes, 'a euphemism for privatisation'.[18] The same is true of DG Markt's complaint that water charges vary hugely from one member state to another and even within member states.[19] But what has this to do with competition? Ireland charges for water through general taxation. Germany, on the other hand, makes a profit from water supply which it transfers to fund other areas of policy. Both systems, and a variety of others, were considered and questioned during the debate leading up to the adoption of the Water Framework Directive and ultimately declared acceptable. As PSIRU asks, does the Commission 'think it possible to create a market which leads to price convergence by enabling people to choose their water supplies from different countries, through water being piped – or perhaps flown? – from Rome to Germany?'[20] Clearly not. What they envisage is rather enabling the big water corporations to compete for contracts against publicly-owned suppliers, which is to say individual local authorities or groups of local authorities.

This was demonstrated most recently when in April 2008 the European Commission opened 'infrigement procedures' against Italy concerning wastewater management services. A group of Italian municipal authorities decided to cooperate in the provision of such services. Normally, the provision of this service by a public authority is exempt from competition rules, but the conditions governing such exemptions are tightly defined. Only an 'in- house' operation, one provided directly by the municipality itself, qualifies. The jointly owned company, Multiservizi, did not fall under the Commission's definition of such a publicly owned service provider, despite the ownership of its shares by a group of municipalities, because no single local authority completely controlled it. The municipalities had clearly set up the company in order to provide an efficient service which each was individually too small to organise. They also provided the service, for a fee, to authorities who were not themselves shareholders. To the Commission, however, this was crucial. The contract should have been put out to tender and not doing so was a breach of public procurement rules easily as important and inflexible as anything brought down Mount Sinai on stone tablets.

As the Commission explained, it

> started infringement proceedings following a complaint by a private waste disposal undertaking. With respect to the waste disposal contracts, the above-mentioned local public authorities are acting as contracting authorities buying services from operators in the market. They can not rely on the so-called 'in-house' exception, because the cooperation structure set up by the parties implies that the municipal companies involved carry out a significant part of their activities for authorities which are not their shareholders. Therefore, under Internal Market rules, the public authorities are obliged to apply transparent and competitive tendering proceedings, opening up the market to competition and ensuring that they get the best value for their citizens' money.[21]

In reality, as Olivier Hoedeman of Corporate Europe Observatory explains:

> the multi-municipal consortium used by the Italian cities is an entirely legitimate and increasingly popular model of organizing public water services, due to its potential to benefit from scale-advantages. This model is widely used, for instance in the Netherlands, France, Germany and Sweden.

This makes the Commission's infringement procedure 'a serious threat to public water delivery in Europe'. Hoedeman, a prime mover behind the founding in 2008 of the European Network for Public Water, calls for 'a moratorium on new EU liberalisation initiatives in the water sector and clear guarantees that public water supply is exempted from EU competition and procurement rules'.[22]

EU Finance

The European Bank for Reconstruction and Development (EBRD), for which the European Community is a principal source of finance, was established in 1991 with the aim of ensuring that the reconstruction of the former Soviet Bloc countries and ex-Yugoslav Republics was conducted on 'market economy' lines. It is therefore unsurprising that it has favoured the private sector when distributing funds. In the first three years of the century, just before seven former Soviet bloc countries and an ex-Yugoslav Republic, Slovenia, became EU members, around two-thirds of the EBRD's grants went to utilities controlled by foreign multinationals. On the other hand, the EU's own long-term financing facility the European Investment Bank, whose activities are not restricted to the EU's own territory or neighbouring countries, has provided funds to back restructuring within the public sector, as has the EU's structural fund, the Instrument for Structural Policy for pre-Accession (ISPA).

ISPA's favouring of the public sector was made necessary by the rules governing European Community pre-accession aid. In order to enable private corporations to live with these rules, the European Commission has developed the concept of the Public–Private Partnership, abandoning its pretended neutrality on the question of ownership structures in favour of enhanced private sector participation. As it states in its document *Guide to Successful Public–Private Partnerships*, the Commission 'has an interest in promoting and developing PPPs within the framework of the grants it provides'. It goes on to warn, however, that 'the use of grants will impose additional conditionalities on projects particularly given the Commission's financing objectives, constraints and over-riding requirement to protect the public interest'.[23]

How well does this document reflect this alleged 'over-riding requirement'? It is, at best, deeply flawed as a discussion of the best way to finance infrastructural development. Although it advises that comparisons be made as to the relative strengths of public and private sector approaches to particular problems, neither here nor anywhere else does the Commission provide any

'Guide to a Successful Public Sector'. Rather than considering the problem of financing infrastructure in genuinely neutral fashion, the Commission has provided 100 pages of advice on involving private corporations in service delivery. This is despite the fact that post-Soviet and pre-accession infrastructural development in central and eastern European countries has involved almost every possible combination of private and public sector initiative, including, as we shall see, some highly successful examples of the latter. Moreover, the Commission has little excuse not to be aware of this, as it financed the Water Time case studies, which included looking at water supply and sanitation systems in Estonia, Lithuania, Romania, Poland and Hungary.[24]

As PSIRU points out, the superiority of the PPP model is 'simply assumed', with a familiar absence of any empirical evidence. Such evidence which is cited comes from studies whose results proved extremely contentious and have been extensively challenged, but this is not mentioned. For the most part, the performance of PPPs is not compared to that of publicly owned entities. Instead, different models of public–private 'partnership' are compared to each other. Protest by opponents of privatisation is counted amongst the 'risks' associated with PPPs, rather than a legitimate democratic activity. In fact, as PSIRU points out, although there is mention of 'the risks of creating a monopoly situation by a private provider', including 'unfair competition... corruption (and) price-fixing',[25] nothing is said about the implication of these 'risks' for consumers or for public sector workers. Worst of all, the fact is ignored that PPPs are possible for major infrastructure undertakings only where contract lengths justify the levels of investment likely to be necessary, a situation fraught with risk. PPP contracts, PSIRU notes, also 'typically involve a loss of transparency and accountability, which increases the long-term risk that the project does not produce the anticipated benefits for the public'.[26]

Negative experiences of privatisation such as those outlined by PSIRU in its report 'Water privatisation and restructuring in Central and Eastern Europe and NIS countries, 2002' (some of which are summarised below) are simply ignored, as are real alternatives.[27]

None of this proves in itself that water supply should not be privatised, only that to do so would create difficulties which would need to be resolved by public authorities. That privatisation ought to be entirely ruled out as a policy approach can, however, be demonstrated empirically, simply by examining recent experience of it in EU member states.

Some Experiences of Privatisation in the EU

England and Wales

Let us begin with the most obvious example, that of the United Kingdom. When the Thatcher government got its hands on Britain's water supply, most of it had been in public ownership for almost a century. The same went for water services, except that they were exclusively in the public domain. Until 1974, when local government was thoroughly reorganised, this meant municipal ownership, with large local authorities taking responsibility for water supply and services and smaller ones forming inter-municipal operating companies.

When local authorities were 'rationalised' into much bigger units in 1974, water was taken away from them. Ten Regional Water Authorities (RWAs) were established, one per river basin authority, each answerable to central government. In the 15 years following the establishment of the RWAs, the workforce was reduced from 80,000 to 50,000.[28]

Typically, the arguments which were used to support the eventual privatisation of water supply and services in 1989 were not backed by any hard evidence. The British people having been subject to a decade-long propaganda campaign to establish in their minds that the private sector was inherently more efficient than the public, none was needed. The RWAs had been starved of funds by a government which sought to reduce public spending to a minimum – though its belief in this, or the application of it, was extremely uneven – and it was easy to create the impression that privatisation would bring much-needed investment capital. Nevertheless, water privatisation remained unpopular. In England and Wales, it was delayed for five years from the original proposed date. In Scotland and Northern Ireland, it was abandoned completely and water in those areas remains in public ownership.

Far from creating a competitive environment, the Water Act of 1988 granted 25-year regional concessions for sanitation and water supply, effectively privatising the existing RWAs and transforming them into private monopolies. In addition, it established 14 smaller companies which were responsible only for water supply, not sewerage. Unlike the ten former RWAs, these were not protected from immediate takeover. They were swiftly snapped up by multinational corporations, further concentrating ownership and reducing any element of 'competition'. The £5 billion debts of the publicly owned companies were written off and the new private entities were given an additional £1.6 billion. The companies were sold off cheaply, moreover, hugely below their market value. Partly because they would have to find substantial sums in order to fund new infrastructure and processes necessitated by EU water legislation, the new companies were allowed to raise prices substantially and were exempt from paying taxes on profits.

The 1980s was the decade during which the British people were cajoled, tricked and bullied into allowing their property to be taken off them and sold for next-to-nothing to people of extraordinary wealth, many of whom had made substantial slices of that wealth available to Thatcher's Conservative party. The snouts were deep in the trough and the trough was filled with wealth created by British working men and women. People were told that they now had, through the sale of shares in the newly privatised companies, the chance to become owners of their own services. The fact that they were already owners of these services was obscured by the Americanisation of an aspect of British thought. People had been persuaded that the state was an entity entirely separate to them, not even potentially under their control and hostile to their interests.[29]

Far from enabling the state to withdraw, privatisation, at least where it establishes privately-owned monopolies, creates a need for strong state agencies to prevent profiteering and other abuses. Britain's water privatisation was indeed accompanied by the setting up of three separate regulatory authorities to cover drinking water quality, pollution of rivers and pricing. How well this last task was performed, if one assumes it was supposed to offer the consumer some protection, can be seen by the fact that pre-tax profits of the ten former RWAs rose by 147 per cent from 1990 to 1998, while real water prices rose by 36 per cent and the real price of sewerage by 42 per cent. Even the pro-Tory *Daily Mail* called the water industry 'the biggest rip-off in Britain'.[30] 11 per cent of water mains and 10 per cent of critical sewers were left in poor condition almost a decade after privatisation. The regulator, Ofwat, was found by a parliamentary committee not to be ensuring that levels of investment were sufficient to maintain the level of service, let alone improve it along the lines promised by the government's pro-privatisation propaganda. In response to Ofwat and the companies' assertion that the level of service was no longer deteriorating, and had not been doing so for five years, the committee was particularly scathing. 'Water companies need to manage and renew their sewers and water mains in order to develop appropriate levels of service to their customers on a sustainable basis.' This was not being done and the 'no-deterioration' approach 'amounted to intellectual neglect of this important problem'.[31]

Companies cut infrastructure investment in order to maintain or increase dividends, with particularly low levels of investment in sewerage. They made much higher levels of profit than did private water companies elsewhere in the world. Vivendi, for example, an enthusiastic international privatiser, recorded profits of 6.3 per cent in 1998, when the lowest level of any UK private water and sewerage provider was 36.5 per cent and the highest almost 60 per cent. Between 1990 and 1998, all bar two firms awarded their senior executives salary increases of between 23 and 200 per cent.[32]

Not all of the vast profits made by the private water companies of England and Wales were given to people who already had more money than is reasonable, however. Some was simply, so to speak, flushed down the toilet. Attempting to enhance their profitability still further, English and Welsh water companies expanded both internationally and into other sectors. In so doing they have been forced to write off over half a billion pounds, their ventures having in most cases turned into loss-making adventures. Companies which were established as debt-free entities by a government which knew how to look after its friends, were by the end of the century back in debt.[33]

Though the biggest losses have come from failed investments overseas, the range of explanations for the waste of the public's money reveals a wide variety of species of incompetence. Operating losses on international water investments, losses as a result of group restructuring, losses on disposal of tangible fixed assets, losses incurred as a result of new 'software solutions', selling off and closure of unsuccessful water businesses in Egypt and in other sectors in the UK, US, Germany and Asia, a golden handshake to a managing director who resigned, problems with a water contract in Bangkok – and so on and so on. You name it, they've lost on it, while all the time their already inflated salaries grew and grew and grew. Meanwhile, the workforce shrunk by over 8500, more than 20 per cent.[34]

Further losses were incurred by the water companies when a public outcry led to a ban on disconnections. Water had always been subject to disconnection if you repeatedly failed to pay your bill. This had been tolerated, for two reasons. Firstly, the publicly owned companies had exercised restraint. Only people who failed to pay their bill on a number of occasions would be targeted. Secondly, it was understood that water was ours. People felt that if the companies took no action against repeated non-payers, other consumers would be in effect paying their bills for them.

As soon as water became something to be sold for a profit, all of this changed. At the very least, the moral picture was clouded. So when privatisation was followed by a sharp rise in the numbers of disconnections, there was an outcry. As the rate of disconnection tripled in the first five years of private ownership, members of parliament began to raise questions in the House related to constituents who had been cut off who clearly did not fall into the 'won't pay' category which the companies had assured the public would be their only target. Some of them, such as a Southampton household with seven children, the youngest aged three, where the mother suffered from a heart condition, sounded as if they had been taken from the pages of the Reports of the Poor Law Commissioners of Victorian times.[35]

The image of the water companies, which appears to have been deadly accurate, quickly became one of callous, profit-greedy conglomerations of the rich, interested in water only because it was the one thing no-one could do without, the ideal lever of blackmail and extortion. And remember, the most vigorous campaign was led not by *Socialist Worker* or even the left-of-centre *Daily Mirror*, but by the solidly right-wing *Daily Mail*.

To object to disconnections it was not actually necessary to feel sympathy for the poor families who were the victims, or even for their children. Cutting off water supplies endangered the health not only of the targeted household but of the broader public. Cases of dysentery were on the increase and doctors and nurses' professional associations were concerned, as were anti-poverty groups.[36]

Put on the spot, the Labour government elected in 1997 acquiesced. No more households would be cut off from the water supply. The companies had responded to public discontent, however, by the installation of pre-payment meters. As the meters worked by cutting off water when not charged with a card – which of course had to be paid for – it was hard for many to see what difference this made to the situation of those unable to pay their bills. The only beneficiaries were the water companies, which no longer had to take a highly public decision and action when cutting off people's water. Ofwat was highly critical of this strategy, while the local authority of Birmingham, England's second biggest city, challenged the meters' legality. As a result of such protests, both disconnections and pre-payment meters were included in the new government's ban on disconnections.[37]

Companies also came under criticism for their failure to deal with the high rate of leakage, which increased the impacts of droughts such as that of 1995. This was again an example of under-investment, where money which should have been spent on improving infrastructure was instead paid out in dividends. But droughts also revealed that under-investment was not the only problem privatisation had either created or failed to solve. Acute water shortages exacerbated the public's lack of confidence in entities which were seen as greedy and self-serving. Whereas in the past people had for the most part respected measures to conserve water, such as hose pipe bans and bans on washing cars, they were ill-inclined to make sacrifices if this meant simply protecting the privatised corporations' ability to pay fat dividends. In 1976 water consumption had fallen by around 25 per cent in response to appeals for restraint in its use. Twenty years later, the effect of such calls was almost non-existent.[38]

The worst affected area was Yorkshire, despite the fact that the west of the historic county is one of England's rainiest densely populated areas. One Yorkshire MP summed up the public mood in a debate on the issue in the House of Commons the following year:

> Yorkshire people are not blind, and they can detect greed and waste. They saw the brimming reservoirs in March, and fished in them. Those reservoirs were full of public water collected from rain in public ownership. They saw and reported leaking pipes and discovered that it took at least two weeks for the company to take action. They read about the 100 million gallons of water leaking daily from company pipes. With one voice – chambers of commerce, Tory and Labour Members of Parliament, local authorities, rich and poor alike and public health bodies – they turned on the company when they were threatened with standpipes, rota cuts, business relocation and constant appeals to restrict their use of water for washing and in the garden. They said, 'No. This is your problem. We, the customers, will not accept the blame.' Indeed, usage started to go up every time a Yorkshire Water boss issued another television appeal.[39]

The quantity of water available was not the only reason people – or 'customers', to use the term customary in the UK – had reason to complain. In one sample taken in 1998, a decade after privatisation, illegally high rates of nitrite, iron, lead and pesticides were found to be present in over 20 per cent of cases. Water-borne parasites have also been a persistent problem. Twenty-five outbreaks of cryptosporidiosis occurred between 1988 and 1998 in England and Wales. Wastewater and water companies were also responsible for one in five pollution incidents during the last ten years of the twentieth century, their first decade of operation.[40]

The privatisation of the water industry in England and Wales was unprecedented and has remained without imitators. In no other case have the whole of the water and sewage systems of any comparable geographic unit been comprehensively privatised. The usual means of privatisation is to grant a concession to a contractor which then operates a water and sewage system which remains in public ownership. In England and Wales, the whole kit and caboodle was given over to corporate capital.

The results, as we have seen, were price increases, a massive rise in operating profits, shareholder dividends and executive salaries, reduction in workforces and low levels of investment. 'Customers' have seen their money used for speculative diver-sifications, many of which have ended up losing large sums of money. In the meantime, ownership became more concentrated in the hands of multinational corporations. Standpipes were seen in towns such as Halifax in the Yorkshire Pennines, where it is said that 'if you can see the hills it's going to rain, if you can't see them it's already raining'.

Since privatisation, driven by new EU laws, the water companies have continued to increase investment. Yet their shareholders paid for almost none of this investment. Instead, it has been financed in part by debt and in large part by government subsidy. None of the promised improvements in efficiency has materialised. Both research and development and on-the-job training have been reduced to the bare minimum.[41]

Finally, privatisation has had hugely detrimental effects on the environment. As a sector, water companies have become the biggest single source of officially recognised 'pollution incidents'. From every point of view bar that of the shareholders and corporate executives who have grown rich on it, privatisation in Britain has been an unmitigated disaster.

'The water industry is no longer run as a public service but as a business whose sole priority is to maximise profits',[42] says Professor Colin Green of Middlesex University in a pamphlet published by an Irish group, Socialist Democracy, which describes itself as 'a Marxist organisation standing in the tradition of Marx, Engels, Lenin, Trotsky and Connolly'. Anything which can provoke the vehement opposition of the Irish far left and the British *Daily Mail*, as well as almost all points between, would surely shake the faith of all but the most fanatical supporters of the 'market'. Yet nothing has been done. The people of England and Wales continue to pay through the nose because they have no choice. The corporations and their executives continue to roll in the proceeds. And the mainstream political parties continue to pretend that all that is needed is a little tinkering with a system which is basically sound.

France

In France, water supply and sanitation services are in principle provided either by the commune – the basic unit of local administration – or by a private company to which the task is delegated by the commune. Many communes are, however, extremely small and these tend to band together to fulfil this responsibility. Large urban communes have also banded together because, as in Paris, separate water services made little sense. France has a long tradition of private ownership of water services, however. The private sector is dominated by massive corporations, while the great majority of publicly owned providers are relatively small. French concerns have tended to centre more on regulation and control than on ownership *per se*.

Privatisation in France has not meant the sale of infrastructure, as it has in England and Wales. Public authorities are required by law to retain ownership, so what they delegate is the actual provision of services and the necessary maintenance and administration which goes with it. In a few cases, communes retained a minority share of ownership, or simply contracted out parts of the service to private operators. For the most part, however, those municipalities which did not retain ownership – which by the end of the last century amounted to well over half of them, covering three-quarters of the country's population – left the operation of services entirely to the private contractors.

Although they are required to maintain ownership of the infrastructure, the municipalities can choose whether and how to delegate control of their services to private companies. At present, 58 per cent of all water supply systems, serving 75 per cent of the population, have contracted out their services. Almost all of them then set performance standards, but the quality of supervision varies and, as we will show below, corruption has been a major problem.[43]

One thing which has undoubtedly fuelled this corruption is the sheer quantity of public money swilling around the system. Water companies, private and public, pay extraction fees to one of six river basin-based agencies. This money is then used to finance infrastructure developments, including dams, flood defences, reservoirs and sewage systems. The communes themselves and both regional and national governments all make further contributions to the agencies' projects. One thing which the French privatisation programme did have in common with the English was that it led to no reduction in the level of subsidies. Private profit, public cost was the rule in both cases, and private companies in France pay only a third of the costs of the infra-structure projects from which they benefit.[44] At the same time, a recent econometric study showed that the kind of Public–Private Partnership, so-called, through which much of France receives its water, delivers a consistently more costly service: allowing for all variables, 'the average price (for delivery of 120 cubic meters of water in a year) jumps from approximately €151 to €176 when a public authority chooses a lease contract instead of managing its own water distribution'.[45]

France is an exception in Europe in having a tradition of private ownership of the water supply and sanitation systems.

Suez Environnement (a subsidiary of the newly created water and energy corporation GDF Suez[46]) and Veolia, 35 per cent of the shares in which are owned by the French state, are by far the biggest private water companies in the world. Together with Saur, a subsidiary of the utility corporation Bouyges, they control 80 per cent of France's water with the rest in public ownership, though the trend is running strongly in favour of this proportion increasing.[47]

Because of this history of private ownership, France's water supply problems have been different from those of newly-privatised England and Wales. In France, concern has long focused primarily on corruption, with water companies providing large donations to both centre-right and centre-left parties and facing a barrage of accusations of bribery, fraudulent accounting and monopolistic practices. Water supply and services are supposedly subject to open tender, but no foreign company has ever been given a look in. There are excellent reasons to keep water in national ownership of one kind or another, but in this case these undoubtedly weighed less heavily than did the large sums of money which flowed (somewhat more reliably than did the water itself) from corrupt corporations to corrupt politicians and corrupt public officials.[48]

What drew attention to these issues were events surrounding the privatisation of the waterworks of the eastern city of Grenoble, which had long been in public ownership. These events were described by the Public Service International Research Unit as 'a text book example of corruption' for which the 'citizens paid the price: increased tariffs and less than honest invoicing methods created profits which, drop by drop, filled the company's coffers'.[49] Alain Carignon, environment minister from 1986 to 1988 and in 1989 the mayor of Grenoble who took the initiative to sell off the city's water services to COGESE, a subsidiary of Lyonnaise des Eaux (LdE), was convicted in 1996 for accepting bribes. Carignon, who at the time had returned to government as minister of communications, was sentenced to five years in prison plus a hefty fine, eventually serving around half of his sentence. LdE executives Jean-Jacques Prompsy and Marc-Michel Merlin were given three and four years respectively, while a lobbyist, Jean-Louis Dutaret, got four years and a fine.[50]

It turned out that the price of privatisation had included a sizeable donation to Carignon's re-election campaign. The privatisation deal was therefore annulled by the courts, but the Grenoble authorities attempted to leave the city's water in the hands of LdE by the simple device of switching to another of the company's subsidiaries. The courts, however, declared this new deal null and void, also cancelling water rates which continued to include the costs of the original, corrupt privatisation. In 2000, the City Council finally bowed to the inevitable and voted to take water back into municipal ownership.[51]

Between 1989 and 1996, in order to cover the costs of their bribes, LdE subjected the people of Grenoble to price increases amounting to FF70m (about €11m) for water supply and FF26m for sewerage. Following the re-negotiation of the contract after the conviction of the two criminals at the heart of this shoddy enterprise, excess income amounted to FF13.7m for water and FF2.3m for sewerage. In the first period, the company invoiced over 51 per cent more water than was consumed, providing gains of FF21m. The post-1996 contract contained an extraordinary provision: if consumption of water decreased, tariffs would go up; if consumption increased, tariffs would go down. A sort of anti-conservation measure, its only purpose was to ensure that the corporation made enough extra cash to get back its bribes. And this, remember, was after the ex-mayor and the LdE executive had been arrested, tried and convicted.[52]

The French system of delegated management of water supply and water services seems to provide an answer to our rather obvious observation above that, as water is a natural local monopoly, privatisation does nothing to introduce competition. If companies must tender to provide water services, then they will be under competitive pressure to produce a good service at a reasonable price. Unfortunately, down here in the real world what the system actually does is provide opportunities for massive corruption. As investigative journalist Julio Godoy discovered when he turned his attention to French water, problems did not begin and end in Grenoble. Rather, the case 'reflected the deep political connections French water companies trade on and the unique nature of the water business itself, where padded billings can be difficult for a customer to detect'. Deception, if not outright fraud, also surrounded the privatisation of water in Paris, with the companies involved hiding enormous profits from the auditors, partly in order to launder away money used to make massive donations to mayor (and later President of the Republic) Jacques Chirac's party. In order to finance this, water bills in the Paris region doubled between 1990 and 2001. In a single year, 1997, one of the companies involved, Suez, took the equivalent of around €350,000 in excess profits from customers who had been deliberately overcharged.[53]

Grenoble and Paris provide only the most spectacular examples of what appears to be rampant corruption in the French water industry. Godoy lists cases leading to prosecutions and prison sentences in Angoulême, where a former mayor got two years, Lyon, where the mayor was forced to resign and eventually received a 15-month suspended sentence, and another case involving 'Vivendi's deputy director general, Jean-Dominique Deschamps, who was found guilty of paying illegal commissions to political parties in exchange for obtaining water contracts in approximately seventy French cities'. Deschamps received an 18-month prison sentence and a substantial fine. Most interesting was the response of one André Fougerousse, mayor of the Strasbourg suburb of Ostwald and member of Strasbourg city council. When forced to resign these positions when illegal payments from three major water companies were detected as having been made to him, including paying for a holiday, Fougerousse admitted the accusations but said that so many elected officials were on the take that it was regarded as normal.[54]

This corruption is not confined to the water industry, but is of a style which particularly characterises abuses at the point in the economy where publicly-funded client meets private supplier. For a long time, David Hall explains, France was plagued by 'systematic, formalised corruption in relation to building contracts'. In Paris, for example, 'the two sides' – builders on the one hand and politicians on the other – 'would meet at a certain time every three months and carve up. The three big building companies were the same, at the time, as the water companies.' Having distributed the contracts, they would 'slice 2% off the price and give it to the political parties in exact proportion to the number of representatives each had on the council. Italy was the same.' It is, Hall insists, 'misleading to think of corruption as a series of individual events involving individual companies and individual politicians'. It is, he says, better described as the work of 'bilateral cartels'.

In general, in France, the neo-liberal drive to privatisation has been tempered not only by fears of more corruption but also by a broad political consensus that national assets must remain in national ownership. In the past, this would not in any sense have meant that they should remain in public ownership. Under EU rules, however, this is the only way to guarantee national control, as otherwise it must be put out to tender across the member states. This means that even if, as we shall see, EU rules undermine many of the advantages of public ownership, support for it remains strong, and not only on the left of the political spectrum. David Hall speculates that this may affect the French water operations of major companies. Saur is already partly in the hands of the publicly owned investment corporation CDC and Suez has merged with Gaz de France (GdF), partly privatising one whilst partly nationalising the other. If French water companies are in part-public ownership, they may be restrained by EU rules for bidding for business abroad. This could lead to them separating their domestic and foreign holdings, establishing separate companies.[55]

France's bad experiences with privatisation, as well as the catastrophic results of privatisation in the UK, have led to a wave of remunicipalisation, as well as to a restraint on the knee-jerk privatisation which appeared until recently to be the likely response throughout Europe to the need to raise capital to upgrade infrastructure. According to David Hall, France's problems 'made a very substantial impact in terms of discrediting privatisation, also beyond France and certainly within it'. He refers to the 1997 report of the Cour des Comptes, the national court of auditors, which revealed major problems in the system: there was a lack of competition exacerbated by a tendency to extend existing contracts without subjecting them to tender and this had created 'substantial profit margins'. All of this was made possible by a lack of transparency and a lack of supervision. This was what had fuelled corruption, including the Grenoble case outlined above, and other examples in Angoulême in the south-west and the overseas département of Réunion. The auditors concluded, in addition, that the companies' claims that rapidly rising prices were a result of heavy infrastructure investment was in many cases untrue. Instead, it was linked to privatisation and the profiteering which came in its wake. Because the water sector was dominated by three massive corporations, there was a major disparity of power in a system which 'left elected councillors on their own, without support, to deal with conglomerates wielding immense political, economic and financial power'.[56]

The reaction to the Cour des Comptes report was, Hall says, 'Enough is enough'. Grenoble had provided 'a text book example of corruption' and though the immediate response was the establishment of a joint venture rather than remunicipalisation, Grenoble's water was back in public ownership shortly after the turn of the century. Hall also points out that since these events 'French critics of privatisation are more active internationally than those from any other country.'

The recently-established 'Water Remunicipalisation Tracker' concerns for the most part developing countries, and we will return to it in Chapter 7, but it also lists seven examples in France, including major cities Paris and Toulouse.[57] It is hard to say whether more will be added, as the battle between the lobbyists of water transnationals on the one hand, with their powerful ideology and simple experience, continues to be fought in the corridors of Brussels and the EU's national capitals. Its results are likely to be drawn out and varied. Four more examples may serve to illustrate why.

Estonia

In Estonia, a relatively poor country which did not become a member state of the EU until 2004, 93 per cent of the urban population is connected to the public water supply. The water sector has recently undergone liberalisation in keeping with the country's obligations as a new member state. In general, however, Estonia has been an enthusiastic privatiser since regaining independence in 1989. The municipalities are caught in the kind of vice which makes privatisation often appear attractive: on the one hand they have been given responsibility for water and wastewater services, on the other they have not been given the resources to fulfil these responsibilities. Whilst in theory they are supposed to operate a full cost recovery system, Estonia remains a relatively poor country in which this is difficult to implement in practice, especially at a time when the EU's legal requirements on water quality demand heavy investment.

Despite these pressures, at the end of the century Estonia had, with the pre-accession help of the municipally owned Stockholm Vatten, managed to retain at least a mixed system, with the capital city's service remaining in municipal hands. However, in January 2001, International Water Limited (IWL), co-owned equally by US and Italian TNCs, formed a joint venture with United Utilities International (UU), a subsidiary of the UK water and energy company United Utilities, to buy a controlling 50.4 per cent stake in Tallinn's water services. The municipality retained the minority interest. According to David Hall, 'Tallinn had a good and successful PUP and then the city council decided they could make money by privatising it.' The company then moved to maximise its profits from the venture. Hall describes this in terms of 'the private company extracting an astonishing amount of money out of the company very quickly. In eighteen months they had taken out an amount equivalent to their entire equity investment, so everything after that was pure profit.' Moreover, the city council did not get exactly what it expected for its money. Hall gives examples of what he calls 'cynical exploitation' such as the fact that 'while the contract specified that it was just for water services the previous company had always done the street gutter cleaning as well. The company started invoicing the council for that on the basis that it wasn't in the contract.' By 2005 water rates had risen by 50 per cent and enormous charges were demanded from the city council for services which had previously been provided at no extra charge. Massive dividends were paid and sizeable payments made to board members for their attendance at meetings. The money for these came from profits accumulated by the company while it was still in public ownership. An agreement was drawn up which would see real prices rise by 54 per cent by 2010. In November 2002, 100 jobs were cut from a total workforce of around 500. This was on top of a 'natural wastage' cut of 100 jobs since privatisation. While the new owners take out vast sums in equity and dividends, the European Bank for Reconstruction and Development (EBRD), a funding instrument financed by EU member state taxpayers, replaces it almost euro for euro. The profits accumulated during the years of public ownership have been given away. Unlike in France, this is not corruption. As was the case in England and Wales, this is simply neo-liberal economics, where private shareholders get the gain while taxpayers get the pain.[58]

Lithuania

In 1994 Stockholm Vatten entered into a twinning arrangement with Lithuanian public sector company Kauno Vandenys. Kaunas, a sizeable city of 430,000 inhabitants, had no wastewater treatment plant, unlike most of the rest of Lithuania. This was addressed through the Kaunas Water and Environment Project (KWEP), co-funded by local sources, the EBRD, the Nordic Environment Finance Corporation (NEFCO), the Finnish and Swedish governments and the EU. These loans were not, of course, unconditional but were attached to financial management criteria, changes in the structure of the company (which, however, left it in fully public ownership), a 20 per cent reduction in staffing levels, a gradual increase in tariffs and comprehensive improvements in financial management, in the effectiveness of bill collection, administration and operations, as well as a comprehensive know-how transfer programme. From 1998 the new company was presided over by a board composed of representatives of the ruling political parties, the city and regional councils, the local energy utility and the city's university.

This model, the Public–Public Partnership, or 'PUP', proved so successful that it has since become the preferred alternative to privatisation in much of the world. According to PSIRU, the 'partners' commitment and the unleashed knowledge transfer appear to lie at the core of the PUP's success'. According to consultants hired to assess the PUP, the company

> faced an enormous pressure to prepare and implement the biggest investment project of its history and, at the same time, transform profoundly its legal status, governance and organisation structure, management systems and practices, customer relations, financial management and information systems. It has succeeded amazingly well, largely due to (Stockholm Vatten) assistance.[59]

The conclusion was that 'The overall impact of the twinning arrangement between (Kauno Vandenys and Stockholm Vatten) is overwhelmingly positive.' With the latter's assistance, the Lithuanian company had 'fundamentally transformed its governance and management systems... (and) adopted modern management approaches faster than most – if not all – other Baltic utilities'.[60]

Along with a comparable scheme in Riga, Latvia,[61] this PUP received positive assessments from a variety of sources. Both PUPs became successful, self-sustaining public enterprises and at the heart of their success was quite clearly the public sector ethos evident both locally and in the foreign publicly-owned company which twinned with them. Yet they outcompeted comparable PPPs at every turn, whether for efficiency, quality, pricing, transparency, staff training or environmental protection. As PSIRU commented, 'Firm reliance on public sector resources on both sides of the partnership, with its not-for-profit basis functioning as a catalyst for effective interaction, have allowed (the partners) to achieve the intended objectives in terms of public interest.'[62]

Sweden

The degree of pressure which municipal and other forms of public ownership of water are under from the neo-liberal ideology currently prevailing within the European Union is best shown by Sweden. Sweden, currently under its second right-wing government of the last seven decades, is a case in point. In the early 1990s, during the first of those right-wing governments, Sweden began a process described by Professor Jan-Erik Gustafsson, a critic of privatisation in his country, as 'outsourcing and full-scale privatisation of the infrastructure sector'. A right-wing coalition won power in Malmö, Sweden's third largest city, for the first time in 40 years. Water services were put out to tender and the contract won by the British firm Anglian Water. The deal fell through, however, and services remained in public ownership. When the first privatisations were carried out, it was by social democrats in Karlskoga and in Norrköping.[63] In an article written for a debate in the magazine *European Business Forum* on private sector involvement in the provision of public services, Gustafsson gives the example of the Norrköping Water and Sanitation Utility, which served a small city of around 120,000 inhabitants. Firstly, in 1997 the public utility was transformed into a corporation and merged with the energy provider to form the Norrköping Environment & Energy Company (NME AB). Two years later the council, run like most of Sweden by the centre-left Social Democrats, Sweden's historically dominant party, went back on an earlier promise and sold part of this entity to a private energy company Sydkraft, a subsidiary of the German energy transnational E.ON, though a controlling 51 per cent stake was initially retained. This was sold in 2001, however, and Sydkraft became sole owner of a now fully-privatised utility. Demands for a referendum on the privatisation were ignored.[64]

According to Gustafsson, the Swedish water sector under public ownership was highly efficient, having reduced staffing levels by 40 per cent, mainly through a process of automation. Charges for water and sanitation were amongst Europe's lowest, despite Sweden's having been at the time one of the continent's most expensive and prosperous countries overall. Gustafsson's point is that it would seem that if one considers the usual arguments for privatisation, it was, at least as far as Sweden went, addressing a problem which did not exist. Even before the election of the 1991–94 conservative government, the Social Democrats had abandoned a socialist perspective on public ownership in favour of a pro-privatisation line which, in Gustafsson's words, proposed that 'core activities of the public sector could be contracted out'.[65]

Once Sweden joined the EU it was therefore primed to swallow the Public–Private Partnership model, the Lisbon process and intrusive competition policies whole. During the Swedish Presidency in 2001, the supposedly social democratic Prime Minister Göran Persson committed the country to 'increased market testing of public services' and 'deepening competition on all local levels'. Pressure for privatisation continues, with the right back in power and Stockholm moving to privatise more of its water services during 2007. In Sweden, local authorities are endowed with a great deal of power, so there is no unified national picture, but, as elsewhere, pressure from EU law is making public ownership ever more difficult to maintain, simply by removing many of its natural advantages.[66]

A similar pattern of decision-making on the local level prevails in Finland, where, however, the preferred model has been to keep water and sanitation in public ownership under groupings of municipalities.[67]

Bulgaria

In December 2000, the EBRD approved a €31 million loan to Sofiiska Voda, a PPP in which the majority owner is the transnational corporation International Water. The funds were supposed to be directed towards infrastructure and service improvements, but much of the money was actually transferred abroad. Other problems arose. It turned out that the company planned to reduce water pressure. People on the upper floors of apartment blocks would have to buy special pumps, or receive no water. This was despite increased prices. In addition, the company was apparently breaking the terms of its contract by contracting out its core activities. Also, the National Water Supplies Trade Union (NWSTU) accused Sofiiska Voda of violating its contract by delegating some of its activities, 'practically turning the concession into a covert privatization', as well as laying off staff, which depart from contractual commitments.[68]

The Picture in the Rest of the EU

French corporations continue to dominate those parts of the European water and sanitation sectors which have fallen into private hands. This pattern, which involves in particular the two water giants Veolia (ex-Vivendi) and Suez, has, however, persisted partly as a result of other TNCs' loss of interest. As we shall see in Chapter 7, this is part of a global trend. What must have looked like easy pickings have turned out to be anything but. Many companies with diverse interests have shed their holdings in water to concentrate on the less problematic profits to be harvested in other sectors. Even Suez and Veolia have been drawing in their horns, though not in Europe.[69]

In a number of countries the public ownership of the water supply is protected by law. In the Netherlands, a law of 2004 requires that household drinking water be supplied by a publicly owned company, but the Public–Private Partnership model has begun to be introduced for wastewater treatment, with the first contract being signed in 2002 by the Delfland water board.[70] In Luxembourg, universal public ownership appears to be unquestioned.[71] In Belgium, strict laws make the country relatively invulnerable to the privatisers.[72] Of course, in many countries where water itself must by law remain in the public domain, such as France, this is not in itself interpreted to mean that its supply cannot be privatised.

In Spain and Italy the involvement of the private sector is longstanding and widespread, and in the latter it is growing. Both countries, as we have seen, face a growing threat of drought and water scarcity, as well as the need to upgrade infrastructure in line with the Water Framework Directive discussed in the next chapter.

Looked at from the point of view of the privatising companies themselves, there seems to have been a marked diminution of enthusiasm since the beginning of the century. Only two transnational corporations, Suez and Veolia, together with their subsidiaries, are heavily represented. Other companies have either dipped their toes in the water and found it to be too chilly for their liking, or withdrawn after making what turned out to be extensive but ill-advised investments. This has increased the market domination of the two French giants, which have maintained the level of their involvement while making adjustments to the details. The withdrawal of other TNCs is, as we shall see in Chapter 7, part of a worldwide pattern. Corporations which had diversified from other sectors into water have withdrawn completely, while many specialist water companies have retreated to their home markets. British companies Anglian Water, Severn Trent and Thames have already withdrawn either partially or fully, or are looking to do so, as have the French group SAUR and German firm Berliner's. Spanish corporation FCC has bucked this trend, but this and other exceptions are too minor to disguise the fact that the problems surrounding investment in water supply and sanitation are putting off those whose only real interest in the matter is its possible profitability. The fact that 'profits' in the sector come almost entirely from either open or disguised public subsidy make it a highly political affair and thus one subject to pressures which may sometimes prove unpredictable.[73]

Conclusion

The EU's determination to make life as difficult as possible for publicly-owned water suppliers has nothing whatsoever to do with 'efficiency' or 'fair competition'. Public ownership can be shown to have no necessary effects on the former, either for good or for bad, while the latter is irrelevant to a sector in which the commodity in question is both essential to life and a natural monopoly. It is, on the contrary, all about lining the pockets of EU-based corporations. David Hall puts it slightly more politely when he says that 'EU policy within the EU and internationally has continued to be characterised by a primary concern to increase market opportunities for corporate business'. When it is suggested to him that such interests are consistently put before those of consumers, workers or the environment, he points out that in the Commission's neo-liberal world-view, 'extending the private sector is in the interests of consumers, workers and the environment. They wouldn't accept the dichotomy.'[74]

Experience, as opposed to ideological prejudice, has demonstrated that private sector ownership of water, or of water and sanitation services, is unnecessary and undesirable. Water is a public good, and can be perfectly well-managed under public ownership and control. That it is not always so managed is no argument for privatisation, but rather one for addressing inefficiencies while maintaining public ownership. Increasing popular involvement in decision-making and democratic control in general and adhering to principles of openness and transparency are essential to this. Structures of ownership of essential services should meet strict criteria based on the general interest and the public good. Such criteria include democratic legitimacy, moral probity and the ability to take both a long term perspective and a broad view. Market economies are not ideally suited to encouraging any of these things. They must be achieved under a system either of public ownership or of strict public control of private operators. The latter is more difficult and more expensive.

Privatisation has fuelled widening wage differentials, leading as it has to huge pay increases for top executives and pay freezes or cuts for the 'lowest' grades, which is to say the men and women who actually do the work. Workers in the public sector, from executives to manual workers, traditionally forgo high salaries in exchange for the satisfaction of serving the public, backed by high job security and decent working conditions. This model is all but dead, but we desperately need to resuscitate it.

Many of the costs associated with privatisation are hidden: the market demands market-based executive salaries, perks and golden handshakes and dividends to shareholders. It needs greater supervision, and, contrary to myth, whole new bureaucracies to manage that supervision. If there is no competition, such supervision must be intense indeed. Where competition does exist, enormous sums are spent on advertising.

A further point, not yet mentioned, is that the public sector has access to much cheaper finance than can be raised by any private company. Privatisation invariably increases the costs of borrowing, unless public institutions do the lending, in which case the soft loan becomes a kind of subsidy. It has been estimated that in replacing private capital with cheaper public finance, renationalisation in England and Wales could lead to savings of £900 million per year.[75] Artificial restrictions on public borrowing, imposed by the rules of the EU's monetary union, have made privatisation more attractive by bringing instant income. Like most quick fixes, however, this turns out to be a decision taken in haste to be regretted at leisure. Lack of money for investment in infrastructure makes privatisation seem the only option. This lack of funds is, however, artificially created.

All of this explains why the invariable effect of privatisation has been an increase in prices. Privatisation also reduces the quality of service, because it is impossible to extract profit from water delivery without doing so: the only way this can be managed is by public subsidy. This is why water is for the most part publicly owned and car manufacture is not. It is clearly possible to make a reasonable profit by making a car, totting up its costs, adding something on top and selling it at that price. Probably most ten-year-olds could explain that that is what making a profit involves, what it means. For water companies, however, it means something quite different. Profits come from public subsidies, whether open or disguised. First you gain control of the most vital of substances, then you use this control to bully public authorities into handing over what are effectively your only profits.

As became evident in the UK when it became the pioneer of neo-liberalism, privatisation is rarely about introducing real markets in which real companies compete for real business. Certainly it is about competition, but this is not a competition to attract customers, not the dynamic competition which its advocates would have us believe is characteristic of 'free market economies'. It is, on the contrary, the competition of the farmyard, a contest of who can dip their snouts deepest into the trough of public money.

6

THE EUROPEAN UNION WITHIN ITS BORDERS, PART 2: THE WATER FRAMEWORK DIRECTIVE

'Look, to be completely honest, you hear it also in the water authorities (in the Netherlands) that the imposition of stringent norms from Europe has done a great deal of good for water quality. Much more than a regional water authority could achieve on its own. A water authority can't ensure that the quality of water is improved as well as when that is imposed from Europe, from above. This has led to enormous improvements. They've forced a number of things to be thoroughly dealt with. A water authority will say, "Look, we've got European standards. We must meet them." And it's the same with the Dutch government. So the investment has to happen, because we're forced to do it. As soon as the local people who run a water authority can decide for themselves whether or not they'll make an investment, they'll ask: "Are they doing this in the rest of the Netherlands?" If you then answer "not if they don't want to," they'll say "then we're not doing it either, because it will put people to too great an expense." So you see that European regulation has given an enormous stimulus, which wouldn't have been felt otherwise.'

Hans Blokland, Member of the European Parliament
for the Dutch Christian Union[1]

Hans Blokland, whose party has in the past been no great enthusiast for the European Union (though since entering the Dutch government as a junior partner it has moderated its stance), perfectly sums up the advantages of European regulation in the area of water quality. Blokland went on to complain about various aspects of EU water regulation which imposed clearly absurd bureaucratic demands on the Netherlands, forcing it to spend time and money on problems which it does not have, simply because the relevant laws had been written in too inflexible a fashion. He also complained that enormous investment would be required in order to bring about quite small and unnecessary improvements in water quality. We would endorse some of these criticisms, but we would also agree that with sound norms in place for most aspects of water quality, we have seen a gradual improvement across the continent during the last three decades. Tapwater is now drinkable. More rivers and lakes can be swum in.

Europe's water problems have changed, however, rather than gone away. Two centuries – more, in some areas – of industrial development have left their mark on our lakes and rivers, increasing the risk and incidence of drought and flood. Less visible forms of pollution than those which once made the air around the Maas or the Irwell scarcely breathable, threaten the health of our environment and its inhabitants, human or animal. For a long time prosperity and population growth increased the amount of water that we use while the amount available remained stable or, in some areas, declined. During the 1990s, these problems came increasingly to the attention of policy makers, resulting in a wholesale change of approach and a new body of legislation.

Since the beginning of the century, water policy within the European Union has been largely organised under the single umbrella of the Water Framework Directive (WFD). The European Community had begun taking action in relation to water in the mid-1970s, but there was a growing feeling at the end of the 1990s that glaring gaps remained to be filled. The aim of the WFD was therefore to address the deficiencies of existing measures and bring almost all water-related provisions into a coherent whole. The prevailing view was that water policy was 'fragmented, in terms both of objectives and of means' and that what was needed was 'a single piece of framework legislation to resolve these problems'.[2]

The Commission's subsequent proposal was in broad terms accepted by the member states. The resulting directive is, for all its weaknesses, a genuinely ground-breaking piece of legislation. What makes it so is the emphasis which it places on water quality, on the place of water in the environment and on sustainability. In addition, the directive's commitment to public consultation and participation in decision-making is, on paper, innovative. Early experience with this, however, was not encouraging. Pieter de Pous of the European Environmental Bureau (EEB), whose members number environmental NGOs large and small throughout Europe, told us that although he hoped the situation would improve as the WFD's obligatory consultation procedures broadened in line with the directive's step-by-step approach, as things stood 'in some places, NGOs don't even bother to answer the questions. It's a waste of time for us. The consultation is not doing anything. It's completely separate from the real decision-making, and it makes no sense for us to look into this.' The process was overly bureaucratic, De Pous complained, and changes were made to plans without any real input from consultations being in evidence. Best practice, he said, was in Scotland and in France, though even in these countries 'it didn't seem that final results from consultation' had any impact on the decisions taken'.[3]

A conference organised by the EEB just before the beginning of the official six-month consultation period in the first half of 2009 heard mixed reports of preparations. A speaker from Baden-Wurtenburg in Germany praised the way in which the matter was being dealt with there, with all levels of government involved in a serious attempt to encourage the participation of NGOs, other interested parties and the broad public. A speaker from Spain told how irrigation-related waste had been reduced by proper, interactive consultation with local farmers and an official of the Scottish government claimed a high degree of success in consulting the public on measures for sustainable flood management. It was clear in general, however, that public consultation is seen by many in authority as an irritation to be handled as perfunctorily as could be got away with and that the further east one goes the more likely this is, in general, to be the case.[4]

The WFD demands both transparency and comparability of hydrological and other relevant data, with the aim of enabling a Europe-wide system for protection of water bodies to be developed and maintained. As Sergiy Moroz of WWF's European Policy Office in Brussels told us, what NGOs want is 'for decisions to be transparent'. An 'innovation' of the WFD 'is to involve the public and encourage active public participation. The idea is that by putting all information on the table and by making very transparent decisions with the involvement of the public, we may actually make better choices about water management.' Yet when we asked Moroz in what way this public involvement had been implemented, he was categorical. 'It hasn't' was his reply, though in qualifying this he was more optimistic. The requirement for consultation and public involvement was 'something very new, or relatively new on this large scale. Many of the member states are learning.' He would clearly like to see them learning more quickly, however. 'This is a problem with the WFD,' he complained. Member states were waiting for deadlines to approach before taking any action at all. 'No one is thinking in advance or what would be beneficial for long-term planning. In the WFD, the first obligatory task is only this year [i.e. 2008]. Public participation until this year was by no means a mandatory obligation. We're only about to see how public participation is actually applied.'[5]

This lagging behind in matters of public consultation becomes still more crucial when one considers the fact that the WFD introduces a clear principle of cost recovery. This could be used to achieve greater equity, but only if combined with a high degree of solidarity and a willingness both to listen to citizens' concerns and to challenge vested interests. Public participation should be fundamental to the construction of any system of cost recovery which seeks to make the polluter pay, or the beneficiary pay when it comes to deciding on where to invest money in improvements. It is also crucial in ensuring that those on low incomes do not suffer and that those who can afford to pay do so without being disproportionately burdened.

The question of 'who pays' is central to the issue not only of equitable charging, but to the effectiveness of the legislation overall. The WFD has built-in socio-economic flexibilities: socio-economic considerations may be used to justify extending the length of time allowed for attaining objectives, for example.

A number of economic instruments are included in order to improve efficiency, including water pricing and cost recovery, and the 'polluter-pays principle'. However, no-one is quite sure as yet how this will work. As Moroz put it, 'at the moment, we're still planning the planning. There's no draft of the river basin management plan on the table where a member state has identified all of these measures.' It was understood, however, that there was a need 'to identify who is causing the pressure, which sector, and ideally, to identify who is currently paying'. Moroz described the process positively, as 'an open and transparent discussion on who should be paying'.

Nevertheless, problems remained when it came to defining what precisely the objectives of this discussion might be. 'First of all,' Moroz pointed out, 'the WFD does not ask for full-cost recovery. It's only asking for adequate contribution of water users and cost recovery of water services.' Which in turn raises a question: what precisely is a water service? Heated debates surrounded this question during the development and adoption process for the directive and the end result is that the WFD is less than clear on this point. Different positions had clearly influenced different sections of the text of the final directive, but as far as WWF and other environmental NGOs were concerned, it was of the greatest importance that the 'polluter-pays principle' should prevail: 'If your sector has been identified as the one that is causing the major impact, if there is an infrastructure serving that particular sector or particular water use, you need to consider that sector a water service and need to identify it as a part of cost recovery.'

The sectors to which Moroz is referring are, primarily, hydropower and navigation. But member states are divided over the inclusion of these in the WFD's scope, as Moroz acknowledged: 'Some of them have actually made a political decision where they say that hydropower is the most important for their economic development, and thus they do not consider hydropower as a water service, and hydropower is therefore not going to be part of cost recovery and redistributing the costs of the implementation of the WFD.'

The implication is that hydropower dams and plant can be built with no regard to any 'external' costs. A hydropower plant is cost-effective if it can produce electricity at a competitive price, in other words, with no regard to any economic, social or environmental costs imposed on the river in question, its surroundings or those who live within its scope.

This becomes an important exclusion when one considers that the WFD also introduces precisely this: the possibility of including environmental and resource costs in the overall balance sheet. 'When you are looking at water pricing and cost recovery, you have to make sure that the costs of damage to the environment are included as well as cost of the forgone opportunities,' Moroz stressed. To take a simple example, if hydropower dams were to be properly costed, a value would have to be placed on the potential product of the agriculture and other productive activities which could have continued on the land which would be flooded when the dam was built. This, Moroz said, is present in the legal text and the legal obligation. The problem is that while it is there it is 'not as strongly worded as we would have hoped', which is why it is possible to exclude hydropower – and other highly intrusive river developments – from any national implementation strategy for the framework directive's vital requirement to include 'external' costs in any assessment. WWF could find only two river basins out of 27 examined, one in Latvia and one in France, where navigation and hydropower had actually been included as water services. Yet an explanatory document published by the European Commission in conjunction with the Common Implementation Strategy does list 'reservoirs for hydropower' as 'water services'.[6]

In July 2006 the EEB went so far as to lodge a complaint with the European Commission regarding the failure of eleven member states to meet the WFD's legal requirements on water pricing policies. The main issue, Pieter de Pous told us, was not just pricing, but the poor quality of economic analysis in general and the restrictive definitions of 'water service' employed in particular. 'The reading by most member states,' De Pous said, 'is that it only includes the service of extracting water for drinking water purposes and treating wastewater to be re-emitted into the environment. These are the classic, traditional water services.' But, he added, they are not inclusive enough to enable the legislation to respond to modern realities. 'We have been arguing that water service is any type of modification or any kind of investment in the infrastructure that you do in order to make use of water, whether it's for drinking, navigation, hydropower, or flood defence, and that means you need to look at all the environmental and resource costs associated with providing the service.'

In the case of agriculture, too, transparency over real costs would be of crucial importance when it came to discussing cost recovery. Both one of the major polluting sectors and the biggest user of water, agriculture receives massive subsidies. 'That can be interpreted as paying agriculture to pollute. No-one is asking member states to give up agriculture, but if you had an adequate public participation in this transparency process, you could start asking questions and start making better decisions.'

As for protecting those who cannot afford to cover the real costs of the water they use, whether domestic consumers or small farmers, Moroz does not believe that their interests, to which he is sympathetic, should be taken into account until after objectives are set. 'You set your objectives as ambitiously as you can. However, when you try to identify the best ways to reach those objectives, that's where affordability can come into question and where it's very clear that for some of the poor families, farmers, and small or medium enterprises, there's no question that you have to look at the affordability. As a state, you are in power to help these people through mechanisms such as water metering or blocked tariffs. The affordability of the public budget can be taken into account. But what we are hearing now from many of the member states is that this is very expensive legislation to implement. Who is going to pay for it?'

Yet, as he pointed out, these same member states committed themselves to the measure. 'There was a reason that they decided that the status of our waters is so bad that we really need to change the way we manage and think about water. Funds must follow.'

De Pous takes a similar line. 'First of all, all cost should be put into the pot or put into the balance sheet as part of transparency,' he argues. Only then should decisions be taken as to who pays and how much they must pay. 'The question of how you are going to share that out between consumers and in consideration of social differences between groups, etc, is a political decision that should be made. We would argue that price should be high enough to save water, so there should be an incentive effect on it. WWF-UK, for example, did a study on block tariffs and how you can adapt tariff structures to protect the poor. There are ways to develop flanking measures to protect socially marginal groups from suffering consequences. Of course this is a good idea, but it's important that costs have to be paid somehow.'

The important issues for environmentalists are, firstly, that sufficient costs are recovered to generate income which can then be used to preserve and improve infrastructure. They stress, however, that governments should be taking political decisions regarding who should be paying. In their view, the polluter should pay, those benefiting from infrastructure improvements should be assessed for their ability to pay and charged accordingly, as should anyone receiving a water supply or service. Decisions on such matters are for the people to take, through their elected representatives, but information must be clear if these decisions are to be effective in achieving stated goals, and if they are to be just. The key word is 'transparency'. If a government simply decides that it is fair or expedient not to charge a certain group the full cost of the water use for which it is responsible, fine, but everyone must know that this decision has been taken and why. 'We are not pushing for full cost recovery,' said de Pous, though even under a full cost recovery system costs could be distributed according to principles of social and economic equity or environmental protection. 'There's always a case for subsidies,' he added, 'but we say that there must be transparency: it's public money and we need to decide what it should be spent on.'

De Pous stresses that this goes much further than simply ensuring affordability for those of limited means. Subsidies might be justified for all kinds of reasons, good and bad. The worst would be that it was simply politically expedient, which is why transparency is so important. Take flood defence. In a country such as the Netherlands, as we saw in Chapter 2, it's clearly needed. As De Pous put it, it would be 'a shame to flood half a country, so for the Netherlands, it's extreme and would make sense to subsidise government policies in order to have sufficient protection'.

The OECD does favour full cost recovery, its Environmental Strategy for the First Decade of the twenty-first century calling for countries to 'Establish policies aimed at recovering the full costs of water services provision and the external costs associated with water use, and provide incentives to use water resources efficiently (demand side management), taking the social impacts of such policies into account.'[7] In the OECD's view, 'the full cost of providing household, industrial and agricultural water services needs to be recovered', the 'polluter pays' and 'user pays' principles applied, and 'a range of economic instruments to provide incentives for efficient water use' applied, 'including by making water prices or charges better reflect the marginal costs of water use, reducing or abolishing subsidies to water use, introducing water abstraction or pollution charges, and using tradable abstraction or pollution permits'. To address the inherent socio-economic inequity of what is, as one would expect from the OECD, an essentially market-based approach, the organisation suggests the use of 'cross-subsidies between different user groups', but also complains that 'OECD households and industries' in some cases 'pay as much as 100 times as much as agricultural users for their water'. Though there are differences of emphasis, the OECD's point is essentially the same as that made by the European NGOs to whose representatives we spoke: it is right and proper to help poor households, small businesses or small farmers and to use subsidies for socio-economic or environmental ends, but this must be done openly and transparently and any financial aid which effectively lowers the cost of water targeted with care.[8]

The directive has been praised for basing water management on the river basin. Whilst this is inarguably the best approach, it was also always one likely to appeal to a transnational body concerned to justify its own existence by demonstrating clear 'value-added'. It requires, and potentially facilitates, international cooperation. It sets a deadline for the achievement of 'good status for all waters', which it defines, and which covers protection of the aquatic environment, including 'specific protection of unique and valuable habitats'. In addition, it sets standards for drinking water and the resources which provide it, and for the protection of bathing water. All surface waters are covered by 'a general requirement for ecological protection, and a general minimum chemical standard'. Groundwater is protected by a ban and a monitoring requirement on direct discharges into such waters, as well as a limit on abstraction fixed at a level which will ensure sustainability. Management of surface water and groundwater are supposed to be integrated.[9]

Sergiy Moroz explained just what the Water Framework Directive would involve for the member states and why, despite all of its weaknesses, it remains the great hope of the European environmentalist movement. For a start, it 'sets very specific obligations on what member states are supposed to do' and the way this needs to work involves genuine problems which have inclined the NGOs to be patient. There are limits, however. The member states, for one thing, had two years to transpose the directive into their national laws, a perfectly normal period but one which seems to have proved too demanding for many member states in this instance. Having transposed the directive, Moroz explained, 'you needed to assess what are the problems with your river basins, to identify main pressures and their impact, which sectors are responsible for that. As part of the characterisation, you needed to look into the economics because the WFD introduces a number of economic instruments. You needed to identify who is paying for water, who is polluting, are these polluters paying? And so on. You need to put these cash flows on the table as part of your state characterization of the river basin.'

This assessment of the status quo was not in itself enough, however, as what was being dealt with was of course a highly dynamic set of systems in a changing situation, including one in which the climate was changing in ways which are, particularly on the local level, hard to predict. So 'you also needed to look at the projected trends, at developments and where they are going, where the water demands will be in the future, what are the potential pressures, like climate change'. In short, 'you really needed to understand where your problems and solutions are, and then the next step is to identify what are the main water management issues you're going to concentrate on'.

This was where the first obligatory consultation should 'kick in'. The idea is that the public would be consulted on 'these significant water management issues' before river basin management plans were drawn up, 'which means that you have objectives to achieve: for example, good ecological status, and a number of parameters in terms of biology, water flow, water temperature, organisms you need to find. You have your objectives set, and your most cost-effective measures identified to reach these objectives.' These draft river basin management plans, which were due by the end of 2008, had then to go into a six month public consultation. 'The very first task is identifying signification water management issues and consulting with the public, and the second step is encouraging public participation in adopting the draft river basin management plans.'

Although the WFD is a framework directive, which means that it is ultimately up to the member states how they implement its requirements, realisation of the complexity of what was involved led to the establishment in 2001 of a Common Implementation Strategy (CIS), which in Moroz's words 'brings to the table representatives of member states who are in charge of the implementation of the WFD, the Commission, and various stakeholders'. This group quickly produced guidance documents on particular aspects of the WFD implementation. 'One of these guidance documents was on what's the best way to arrange public participation. One of the key principles in there is that you do it locally. This is the requirement of the directive. All the EU territories are divided into river basin districts, and there is a river basin authority that is in charge. The idea is that it's the people who are affected by these decisions who need to be consulted. We have a number of good examples where this has happened where there were quite extensive consultations that went through the local community. People were drawing maps and identifying particularly what they want to do in the local river basin, what the priorities are, and how they want to see their water managed.' There were, however, 'also very bad examples of public participation'. The worst were in a number of the new member states, where traditions of public consultation simply did not exist and 'involving the public in political decisions or allowing NGOs access even to background information documents to engage in meaningful discussion' was alien. They were, Moroz feels, 'learning' – once more, that optimistic word – including 'the value that NGOs can bring to the table'. The NGOs had high hopes of the Water Framework Directive and had not been willing to let this, in their view crucial, aspect of its requirements slide. And the news has been by no means universally bad. Regular monitoring has shown that 'one of the areas where there are some improvements is public participation. Again, it's by no means adequate, and there are still shortcomings.'

It is worth listing the principal elements of what constitutes a River Basin Management Plan, as these provide the foundation of water policy within the European Union and are likely to do so for some time. The WFD requires that they include the mapping and identification of protected areas and a map showing monitoring results for surface water, groundwater and protected areas. They are then required to list environmental objectives and to make an economic analysis of water use. As the Directive requires that the public be informed and consulted, details of this must be included in the plan, including any changes which have been made as a result of such consultation. Exactly what measures will be taken in order to fulfil the WFD's requirements for water protection and cost recovery must be detailed. Controls on abstraction designed to meet the sustainability requirement must also be included, as well as controls on discharges and other activities which might have an impact on water quality. Direct discharges to groundwater, which are allowed only in exceptional circumstances, must therefore be identified, as must measures taken to prevent the most serious forms of chemical pollution, to prevent or reduce the impact of accidental pollution and measures taken to address the problems of water bodies which do not seem likely to achieve good status within the timescale set. A range of other details relating to pollution is also specified, including a requirement for a list of any human activities which might have an impact on surface waters or groundwater, including through both point source pollution and diffuse source pollution.[10] Pieter de

Pous sees this last as 'one of the main problems' and as one of its main sources is agriculture, it cannot be properly tackled without major changes in a CAP which does nothing to discourage the use of chemically polluting fertilisers and pesticides. Nevertheless, potentially effective measures – the Nitrates Directive and the Waste Water Directive – actually predate the WFD. The problem has been that implementation has been inconsistent. In 1992 the EEB declared itself 'pleased with the Nitrates Directive, though not with the implementation record of most Member States' and little seems to have improved since.[11] Pieter de Pous describes nitrate pollution as 'an old problem that we haven't been able to properly deal with' adding that 'we have the Nitrates Directive and Waste Water Directive: with all their policies and instruments, if we implement this and deal with that, we deal with a large part of the problems'.

In 2001, the EEB reported that the problem was not inadequacies in the legislation so much as the fact that 'no directive has been completely implemented and applied by the Member States... The implementation situation may be called disastrous, and in terms of EU-wide common water protection standards, the Community is far from its goal.' Little has changed. Despite the establishment of IMPEL, the new Network for Environmental Inspection, the '(e)fficient, effective and realistic solutions' for which the EEB called are still not in sight.[12] A survey conducted by the European Commission in 2004 revealed that the pattern of 'serious shortcomings in the implementation of EU environmental law' continued. Member states were guilty of late transposition of directives, incorrect transposition and failing to meet all of a directive's obligations. The worst offenders were 'France, Greece, Ireland, Italy and Spain' and water and waste were among the worst sectors. Indeed, in 2001 one researcher concluded that if it weren't for a small number of offenders – a slightly different list this, featuring Italy, Greece, France and Belgium – there would be no serious compliance problem.[13]

Although these countries remain among the worst for implementation, things have clearly deteriorated since then, however. At the end of 2003, the environment sector was responsible for more than a third of ongoing infringement procedures, amounting to over 500 individual cases.[14] Moreover, no significant improvement had been registered by 2007, when 479 infringement cases were ongoing (though this was admittedly now only 22 per cent of the total across all legislative areas). The Commission even proposed making serious environmental offences subject to criminal sanctions throughout the Union, though this was regarded by many as simply Commission empire building: traditionally, member states have jealously guarded control of their criminal law codes, quite rightly seeing this as a *sine qua non* of sovereignty.[15]

Clearly this proposal also reflects a certain impatience, which is certainly shared by the NGOs involved. Weaknesses in a directive itself are one thing, but there is in many ways a stronger reaction when a law is won and its valuable aspects are not implemented or enforced. The WFD, in De Pous' words, is 'full of ambition and goodwill'. He sees it as important to remind member states that they had agreed to the policies in the directive 'because we needed an ambitious approach to dealing with water problems'. Now, however, 'what you see is that countries have all sorts of ways of looking for emergency exits. They come up with their plan, which, to a large extent, consists of things they were doing anyway.' Rather than looking at what is needed to meet the objectives of the WFD, there is a tendency for countries to try to pass their existing policies off as adequate to the task. Because of such backsliding, what NGOs and others sympathetic to the directive's aims have to do 'is to keep up regulatory pressure'. The member states 'signed up to something; it's not just something imposed by Brussels but what they collectively agreed to in 2000. We have a common market, etc., so we want everyone to play by the same environmental rules.'

The possibility of the threat of court action remains as a potential weapon in the armoury, though it was not the only such weapon. Just as important, De Pous insists, and the focus of the NGOs' work in this crucial phase of implementation, is to engage the general public, especially the people who live next to the water in question, to get them involved. The presentation of the draft River Basin Management Plans at the end of 2008 will be followed by a statutory six-month consultation period. This must be put to good use. De Pous, himself a Dutchman, gives the example of algal blooms in the Netherlands, a result of eutrophication, where an excess of nutrients from agricultural runoff provokes the growth of these organisms at the cost of other flora and fauna. Not only does this threaten biodiversity but the algae also pose a threat to human health. The result is that 'you can't swim in a large number of lakes any more. There's a very good reason to do something about it, and it could be done by 2015 (the target date in the WFD for good water quality). We tell the government that it's their job to deliver.'

In the UK, in an approach of which De Pous clearly approves, NGOs have produced a blueprint for water: ten steps to sustainable water by 2015. De Pous explains that the 'blueprint' identifies detailed problems before suggesting solutions and that it mentions the WFD only once.[16] The EU has never been popular in Britain so, De Pous says, 'Not mentioning Brussels is the best way. It's increasingly the same in the Netherlands and probably in more countries.' The point is that the popularity or otherwise of the EU should be irrelevant. The directive was negotiated by the then 15 member states and has been signed up to by the 12 which have joined since. A measure of this kind could, in theory, have been negotiated between 27 fully independent countries. It is not a matter of being 'pro-' or 'anti-European'. The measures contained in the WFD are urgently needed.

The OECD has praised the Water Framework Directive as being precisely the sort of 'strong legal framework' based on 'integrated water resources management' which it would like to see adopted everywhere.[17] It is certainly the case that this is what it set out to be, and it is also true that even critics have many positive things to say about it. However, apart from the clear and continuing problem of implementation, the legislation itself is far from perfect. Sergiy Moroz, though generally positive about the directive, calls it 'a compromised text', explaining that it was the result of 'a very tough negotiation' and adding that the timing of this negotiation had been fortuitous. Environmental legislation was 'politically major', perhaps because the centre-left was in the ascendancy amongst the majority of member state governments. In Moroz's view, 'we would never get anything like the WFD in the current political climate'.[18] Though he did not say so, this presumably means that interpretation of the directive will be laxer, its implementation less enthusiastic than might otherwise have been the case. Moroz nevertheless stresses that WWF believes that the measure can 'deliver'.

He is particularly enthusiastic about the fact that, as he puts it, 'for the very first time, it tries to put ecology into the law'. The way that he talks about the Directive makes it clear that he identifies strongly with it: 'We're trying to look at it in an integrated manner and not just trying to solve the end-of-pipe solutions, which was in the previous water legislation. We really try to look at all the water users in all the ways we manage, protect or abuse, conserve water, and try to set measurable targets, which turns out to be a very difficult part.'

Despite all the problems, Moroz believes that 'the potential of the WFD is huge'. However, 'because it's so ambitious and requires member states to change business-as-usual and start thinking about water differently, that's what we see happening on the ground when the WFD is being implemented: the reluctance to change business-as-usual, the reluctance to embrace new thinking'. He identifies agriculture as the major sector standing in the way of effective implementation, acknowledging that 'there are very tough political choices to be made'.

The European Environmental Bureau (EEB) recognised the limitations of the Water Framework Directive from the word go, criticising it for offering 'complicated and wide-ranging exemption and derogation conditions for the environmental objectives', for creating legal uncertainties and for pushing back important decisions, such as the criteria for assessing groundwater quality, to a later date.[19] It went on to complain of '(l)ong deadlines, ambiguous provisions, an unclear level of protection as well as a large number of opt-out clauses and time extensions'. Nevertheless, its conclusion was that the WFD had 'sufficient potential to have an overall positive impact on water resource management in Europe'. The question is, then, has the Directive fulfilled that potential, or does it at least look like doing so?

Reports from the NGOs are mixed, but they are clearly putting a brave face on matters. In 2001 the EEB described the Water Framework Directive as 'a major step towards sustainable water management', citing in particular its aim to achieve good ecological and chemical status by 2015. In 2005, however, a risk assessment carried out throughout the EU-27 concluded that a significant number of surface and groundwater bodies were at risk of not achieving good status by 2015.[20]

Now they say that in many cases this status will not be achieved. When we put this to him, Pieter de Pous, Water Policy Officer for the EEB, admits that 'the more enthusiastic quotes on the WFD come from a time when we were deeply involved in negotiations. They ended at 4 a.m! On the one hand, we had some serious disappointments in the final results, but on the other the basic concepts of the river basin management approach – the ecological objectives in water management – got everybody in the environmental movement very excited.' Now, talking to a number of people from environmental NGOs, that excitement is clearly dissipated, though there remains a strong feeling that the WFD at least gives us something to work with.

There is a continuing problem of chemical pollution. Though effective measures have been taken which have addressed aspects of this, there is a huge grey area of unknowns and, in the face of the enormous financial interests and lobbying powers of the industries most responsible for chemical pollution, an at best patchy application of the supposed EU fundamentals of 'the polluter pays' and the 'precautionary principle'. In addition, the legislative process leading to the WFD cast light on a number of other sources of water degradation, which the EEB and WWF list as 'the physical deterioration of water habitats through abstractions, dams and dykes which serve electricity generation, transport, water supply and flood management to such an extent that aquatic ecosystems are significantly damaged'.[21]

The question of deadlines and the failure to meet them covers more than the achievement of 'good status', the condition which member states must aim for in dealing with rivers, lakes and groundwater. The WFD sets deadlines for the implementation of almost all of its various provisions. The River Basin Management Plans, which are absolutely central to the achievement of the Directive's stated goals, must be presented by December 2008, operational by 2012 and have fully delivered their environmental objectives by 2015. In its interim report of 2007, the Commission was clearly already preparing the ground for the failure to meet these targets.

Firstly, the report says the starting point turned out to have been 'worse than expected', with the 'percentage of water bodies meeting all the WFD objectives...low, in some Member States as low as 1%', though 'the results need to be analysed in more detail'.[22] Secondly, legislation pre-dating the Water Framework Directive, covering for example waste water discharges, nutrient runoff from farmland, industrial emissions and discharge of hazardous substances, had clearly been unevenly implemented. For new member states, transitional periods for implementation were of course still running and in most cases would last until 2015.[23] Thirdly, not only had the Commission been obliged to take legal action against eleven of the then 15 member states for failing to transpose the WFD's requirements by the deadline of December 2003, but when transposition was finally enacted its quality had been 'poor'.

The Commission did declare itself more or less satisfied with the administrative arrangements established by the member states in regard to river basin districts, though many of these had not been completed by the deadline. In addition, the Commission noted that 'actual performance will only become evident in practice over the coming years'.[24]

Further complaints include the unevenness of quality of the member states' environmental assessment reports, which are required by the Directive. Though a number produced 'a good or satisfactory report', the unevenness and the fact that in every instance 'data gaps need to be filled in order to provide a solid basis for the 2009 river basin management plans' would make monitoring difficult. As for economic analysis, the weakness of this was the reason why a number of reports 'clearly do not meet the minimum requirements of the Directive'. Particular concerns were 'the proper identification of water services and uses, and the assessment of cost-recovery'.[25]

A statutory consultation process to run from the presentation of the member states' River Basin Management Plans at the beginning of 2009 to the mid-point of the year required each country to identify 'Significant Water Management Issues', which in turn meant 'developing priorities and deciding on measures and objectives' for these plans. These were expected to concern, above all, '(h)ydromorphological deterioration and alterations' and the need to respond to these through 'economic instruments to reduce water consumption' as well as 'planning regulations and river restoration programmes to increase the connectivity between land, groundwater and rivers'. NGOs were concerned that more and more waters were being listed as 'heavily modified', a category which allows less stringent objectives to apply. They were also worried about the fact that 'key data on biological elements to support coherent and WFD-compliant ecological classification is missing'. This may sound complicated, but all it really means is that member states were cheating. Instead of deciding what the environmental problems involved in achieving good ecological status were and then debating how the resources might be found to tackle them, they were deciding how much they were willing to spend and then tailoring their definition of such status accordingly. Economic instruments were generally discounted as a means of managing water.[26] Sometimes this may be for fear of increasing the financial burden on poor households, a worthy aim but one which can surely be achieved within an overall programme designed to discourage the wasteful use and abuse of water. More often it is because the lobbying power of the major polluters is so great that they are able to evade responsibility.

Drought and Water Scarcity

A further area of concern to NGOs was the way in which the WFD had been negotiated with little consideration for climate change.

As Sergiy Moroz conceded, 'climate change started to dominate the agenda quite recently. The WFD was drafted with no particular mention of it.' On the other hand 'there are definitely spaces where climate change can enter into the river basin management plans, and the Commission's claim that the Water Framework Directive provides a consistent framework for integrated water resources management may still be true'. The Commission itself notes that the directive 'doesn't address the matter directly', expressing the hope that the end of 2009, when the implementation plans for the first cycle must be presented, will put this to rights.

This is also Moroz's hope. He agrees that climate change is not excluded by the projected structure of the river basin management plans. What the WWF and other concerned NGOs are trying to promote is, he says, 'that you don't lose time' and therefore 'in the first river basin management cycle, you try to make sure that all the measures and investments you make are climate-proof. You make your decisions now and think long-term on the likely impacts of climate change.' He acknowledges there are problems, especially as, while it's easy to talk about the global level of climate change and the IPCC breaks this down into continents, information on the local and regional effects is scant and patchy. Stronger action, he says, may have to wait for the second cycle, which begins in 2012. So that while the Commission argues that 'efficient pricing policies, making water-saving a priority and improving efficiency in all sectors are already essential elements of the EU's approach', this has yet to feed into the member states' policies under the WFD.[27]

In the NGOs' view, the WFD provides a necessary instrument, a tool for addressing climate change which should not be passed up. Its approach, Moroz argues, is potentially the 'best adaptation strategy to climate change: increasing the natural resilience of the ecosystems, working with nature when you manage water... Nature can adapt. Functioning deltas can rise with the sea level. If you have a functioning flood plain on the river during the flood peaks, it will have space to spread; the wetlands will store it and release it gradually so the flood does not have such a big impact. Same with drought, which is a natural event. If you have a functioning natural ecosystem, it is perfectly capable of coping with drought. The idea is that your best adaptation strategy is to try to increase the natural resilience of the ecosystem. Ambitious implementation of the WFD goes directly in that direction.'

Unfortunately, many EU policies are simply not integrated with some of its attempts to address climate change and this is 'one of the main problems of why the WFD is not succeeding. Many of the responses to climate change mitigation are counter-productive.'

There is a push for hydropower in many member states, but not nearly enough is being done to modernise infrastructure to make it energy efficient. River transport is being promoted as a climate-friendly alternative, but the result is that 'we're looking at navigation and seeing all the rivers as canals, such as the lower Danube in which the last stretch is a perfectly functioning river with the last migrating space for sturgeons, which is being threatened by EU-funded plans to improve navigation'.

Moroz denies that he is opposed to water transport, but he wants, he says, to see money invested in infrastructure so that 'navigation and ecological needs can co-exist. You can invest the money for infrastructure into modern technology. These ships exist that are wider and take the same amount of cargo, but they don't require the same depth so you don't need to do any river training works such as dredging or bank enforcement. You can use it as it is: the environment benefits and economic need benefits. You really need to take both into account and try to find a compromise.'

Pieter de Pous warns that where navigation takes priority, the environment will in many cases suffer. 'The problem,' he explains 'is that rivers like the Danube and the Elbe are continuously being worked on and deepened so that sufficient ships can go through.' Take the example, in particular, of the Danube: 'It's a rain-fed river, so that the water level varies enormously throughout the year. In order to have sufficient water, you need to dredge the lot and build dams. So the river becomes a series of bath tubs, and you need to go from one bath tub to another to move around. If you want to keep navigation for the Danube in the long run, it's a massive investment if you only look at the financial cost. If you look at the resource cost and the environmental cost, then you have completely disturbed the ecosystem.' So what happens to the idea of a fully commercially navigable Danube? 'We would obviously say that this is not worth it. Other possibilities are railway transport or the idea of decoupling transport from economic growth, which has been forgotten these days. We want the cost from all these services to be visible, transparent political decisions on who will pay for it, and a public discussion on whether it's fair to do it or why we wouldn't want to do it.'

Similarly, De Pous fears that climate change will give a boost to another technological 'solution' incompatible with good river management: dams. Much better, he says, to adapt water management to climate change. Like Moroz, he wants to see an 'ecosystem approach' designed to enhance 'system resilience': 'If you have clean water, you don't need to spend as much energy to clean it up. If you have full ground water aquifers, you don't need to build reservoirs with high evaporation and pollution risks to get your water. If you have natural forest and vegetation and good soil quality, your groundwater recharges more quickly, and you get less run-off and fewer erosion problems.'

These, he says, 'are cost-effective and simple ways to actually adapt to climate change, which, in many cases, are not compatible with mitigations being propagated at this point, like biofuels, hydropower, and investments in infrastructure. This hasn't really entered into thinking.' He fears that many member states are waiting until 2015 or even 2021 to 'see where we are' before considering how to adjust their water management. This, he says, 'is a big mistake, because over the next six years we will see massive impacts on rivers and water bodies, which will adversely affect our adaptation possibilities. It will be too late.'

De Pous also worries about the recent tendency to use climate change as propaganda for a resurgence of nuclear power. Apart from the possibility of accidents and the difficulties posed by storing waste that will remain deadly poisonous for thousands of years, there are the direct consequences for water courses brought about by the amount of water for cooling needed to service a nuclear power plant. He points out that in France during 2006 reactors had to be closed because of a shortage of cooling water. Using water in this way means that it returns to the river warmer, with 'serious impacts'.

In its Green Paper on the issue, the Commission pays lip service to the kind of holistic, cross-sectoral approach demanded by climate change, under which no decisions would be taken in one sector without assessing their broader impact.[28] In policy terms, however, there is very little evidence of such mainstreaming. As we cannot know exactly what climate change will bring, we should be making assumptions based on the precautionary principle and planning accordingly. This planning should involve not only water itself or water services, but also anything which might affect agriculture, urban and industrial demand management, or the broader environment. Only economic planning can combat climate change and the only answer to those who associate a planned economy with economic stagnation is a high degree of public involvement in decision-making.

Peter Gleick, in an analysis of the problems involved in designing water policy in the face of climate change written over a decade ago but more resonant than ever, gave a concise account of the issues. 'A lot is at stake', Gleick said:

> Water-related infrastructure is extremely expensive to design and build, and it lasts for many years. Systems for storing and delivering clean drinking water and for collecting and disposing of human wastes have largely eliminated water-related diseases in the industrialized world. A substantial amount of our electricity is generated by hydropower facilities. Productive agriculture in many parts of the world depends on reliable irrigation flows. And extreme hydrological events, including both floods and droughts, wreak havoc on lives and property despite our best efforts to prevent them.[29]

Gleick's conclusion is that the only way to proceed is through what he calls a 'no regrets' approach 'which includes evaluating management and operational options under a broader range of climate scenarios than managers usually think about'. The problem is that until now:

> Water planning and management relied on the assumption that future climactic conditions would be the same as past conditions, and all our water-supply systems were designed with this assumption in mind. Dams are sized and built using available information on existing flows in rivers and the size and frequency of expected floods and droughts. Reservoirs are operated for multiple purposes using the past hydrologic record to guide decisions. Irrigation systems are designed using historical information on temperature, water availability, and soil water requirements.[30]

The problem is, however, not so much that this infrastructure will no longer be well-adapted to our needs, but that it will be impossible simply to adjust this. The difficulty is therefore one of uncertainty. If we could predict exactly what climate change would entail for each specific river basin, dams, reservoirs, flood defences and other infrastructure could be adjusted accordingly. This might still not be the best option, but it would at least be available. As climate change is an inherently unpredictable process, however, above all on the local or regional scale, this option is in any case simply not available. As Gleick says, any such approach 'runs the risk of making expensive, incorrect design decisions'.[31]

And yet that is precisely what the EU, through its funding, is encouraging the member states to do. Spain is addressing its water problems through extensive dam-building, and if you're a taxpayer anywhere in the European Union you're helping to fund a series of projects which are environmentally destructive and massively disturbing to aquatic and surrounding ecosystems. WWF gives two examples which have implications for its campaign to save the threatened Iberian lynx. For La Brena II dam, the European Regional Development Fund provided almost €80 million and for Arenoso Dam almost €30 million, in both cases about half of the dam's total cost.[32]

Hydropower, seen as a way of mitigating climate change, is a driving force behind dam construction which has nothing to do with water supply and, as we have seen, often has conflicting priorities. De Pous: 'Hydropower has been with us for a very long time, but you see a current hydropower revival and it's very difficult to argue with it because with climate change, we have to do it. Of course, there are a lot of laxities around it; there are a lot of things you can do about climate change without going into the last bits of river that we still have in a natural state and changing that as well. One thing you can do is replace a lot of the existing turbines and existing dams, with a gain of 30% output in energy. There's quite a lot you can do with existing dams.'

Although there is some truth in the idea that the major focus of dam-building has switched to developing countries, a total of eight new dams is planned for the Danube river basin alone, an area which covers member states Germany, Austria, Slovakia, Hungary, Romania and Bulgaria, as well as five countries outside the EU. Spain is planning to add to its 1200 dams and Portugal and Poland are each planning new constructions. Despite all the evidence, there is a widespread conviction that dams are an answer to the increasing incidence of drought which will accompany climate change. The reasons for this are clear: while they may make life harder for the rest of us, they are of immediate and short-term benefit to the agribusiness corporations which want to see them built and these exert more influence in Brussels than do environmentalist arguments, the citizens of the member states or scientific evidence.

Despite the cancellation of some of its most damaging elements by the centre-left government which replaced the right in 2004, the Spanish National Hydrological Plan (SNHP) foresees the construction of 118 dams and the re-routing of 22 rivers. Yet Spain relies on hopelessly antiquated, inefficient systems of irrigation which could be updated at a fraction of the cost of these mega-projects.[33] Other options include switching to less thirsty crops – something actually required by the WFD in drought-prone areas, though it is not clear whether or how this will be enforced – or reusing urban wastewater. All of this would cost money, of course, but it would surely be a much better use of EU taxpayers' cash than would a series of environmentally destructive dams.[34]

Pieter de Pous is quite clear: dam-building is part of the problem, not part of the solution. 'If you don't maintain your environmental flows, in southern countries the already existing problem will get worse,' he says. And dams don't help. Neither do other mega-engineering projects, some of which are being justified, as we note above, as measures to mitigate climate change. 'The big upcoming problem is hydromorphological pressure, which is the engineering of the river. It is the shaping of the river for navigation, dredging, deepening the river, putting in weirs and sluices to keep the water level high enough, for enough spaces for the boats to go through, and so on.' This would in many cases, De Pous says, be fantastically expensive, but it would also be hugely environmentally destructive, because of the need for 'a sufficient amount of space for natural processes to take place, for ecosystems to function, to provide services and to act as a buffer'.

Instead of these technological solutions, De Pous argues we should be working to reduce demand. 'If you give people the impression that there is sufficient water forever, you won't be successful. The main thing is to radically bring down demand, and the Commission estimates that we can bring it down 40% in the EU, cross sector. Focus should be on demand management, then supply management.'

Austrian Christian Democrat MEP Richard Seeber, the European Parliament's Rapporteur for its response to the Commission's proposals for dealing with water scarcity, argues that each Directorate-General of the Commission should 'adopt a short-term plan on water scarcity related to their particular policy fields. There, they could identify actions that need to be taken. There are a lot of opportunities we find when we search the EU level.'[35]

The proposals which Seeber was addressing were contained in a 2007 'Communication from the Commission'.[36] A Communication is a set of musings, a discussion document. And this one certainly sparked off discussion. The EEB declared its support for 'most parts of the Commission's proposed strategy', praising its prioritising of demand management over increasing supply, which should be a 'last resort'. Predictably, however, the policies put forward were lacking in one crucial respect: there was a 'lack of clear policy options to address water-use in agriculture'. As the EEB pointed out, agriculture is responsible for an average of almost 70 per cent of water consumption in the EU, most of it for irrigation. In many countries water is provided to farmers free of charge, so that it is a common sight to see fields being sprayed continuously by sprinklers in the full heat of a summer afternoon. Irrigation continues to be aided by EU subsidies. No policy options are formulated to deal with this. The EEB wants to see money now used on subsidising irrigation spent instead on water saving, but it is hard to see how we get from where we are now to where the EEB would like us to be.[37]

Flood

From just before the end of the last century, Europe, in common with many other parts of the world, has suffered a series of frequent and often catastrophic major floods, with the death toll reaching over 700 in the decade from 1998. In addition, half a million people have been displaced and €25 billion of insured property destroyed.[38]

This has prompted action at European level and towards the end of 2007 the EU introduced a new directive on 'the assessment and management of flood risks'. Member states were required by the directive to assess water courses and coastlines to estimate the risk of flooding in order 'to reduce and manage the risks that floods pose to human health, the environment, cultural heritage and economic activity'. Flood risk management plans must be coordinated to the river basin management plans required by the WFD. All relevant materials must be made available to the public and public consultation procedures required by the two directives coordinated. Where river basins are shared by two or more member states (or by one or more member states with non-member states), nothing should be done which would increase the risk of flooding to the other states involved. Plans are also required to 'take into consideration long term developments, including climate change, as well as sustainable land use practices'.[39]

The EU's own official research body, the European Environmental Agency (EEA), lists the potential environmental and public health effects of different kinds of floods as blockage of water treatment plants (and subsequent release of contaminants), damage to vegetation, including crops, 'the mobilisation of contaminants present in the soil', for example in the proximity of chemical plants, soil erosion, the infiltration of polluted runoff into aquifers and the breakdown of drainage systems. In warmer areas soil erosion caused by flooding can combine with forest fires and soil degradation to cause desertification. The EEA speculates that, apart from the effects of climate change, it is changes in land use, particularly urbanisation and infrastructure development, which have increased Europe's incidence of flood. During the last two decades of the twentieth century urban development increased by 20 per cent in Europe, though population overall went up by only 6 per cent. Much of this development has taken place in areas vulnerable to flood.[40]

The European Commission's proposals for dealing with the threat of flood were heavily criticised by environmentalist NGOs who accused the Commission of ignoring the scientific consensus and choosing outmoded methods of flood control. Outmoded they may be, but they are also profitable for the construction companies which employ hundreds of lobbyists in Brussels and potentially prestigious for the political leaders making the ultimate decisions.

When the European Parliament declined to amend substantially the Commission's proposals, despite their failure to promote the sustainable management of floods, the European Environmental Bureau, Friends of the Earth Europe and the WWF accused MEPs of favouring an unrealistic 'business as usual' approach which fell well short of what was needed. Instead, they argued, the EU should have been developing a system which 'works with natural defences like wetlands, floodplains and riverbank woodlands'. What Europe was left with was an old-fashioned approach based on 'man-made concrete structures to constrain flooding'. Persisting with these practices would 'give rise to ever-worsening impacts on society, the economy and biodiversity'. What these NGOs wanted to see was the re-creation of natural floodplains and the internalisation of external costs, under which environmental and other hidden costs would be calculated in to arrive at the real price of any development. Systems implied by the new flood management directive would not, in effect, be required to comply fully with the requirements of the WFD, whose binding objectives were to achieve 'good water status' and 'prevent deterioration' of water bodies.[41] As we saw in Chapter 2, the NGOs' criticisms were entirely in keeping with current scientific and technological knowledge. The Commission, putting forward weak, misguided proposals under pressure from 'industry', must have known that this was so.

The Flood Directive attempted to provide an EU-wide coordinated system comprising three elements: a preliminary flood risk assessment, identification, classification (according to seriousness of risk) and mapping of flood risk areas, and thirdly flood risk management plans which would begin by attempting prevention. Each of these stages was given a binding target date: 2011, 2013 and 2015 respectively. Yet despite these legally enforceable target dates (and unlike in the case of the WFD) environmentalist groups had little positive to say, arguing that it would conflict with some of the framework directive's most positive aspects. The differing water management processes envisaged by the two measures, they warned, would produce confusion and unnecessary bureaucratic duplication.

The Flood Directive, they argued, would conflict with the WFD in terms of the water management processes envisaged by the two measures, producing confusion and unnecessary bureaucratic duplication. To add to this was a vagueness of wording which made it unclear whether all measures taken under the Flood Directive would be required to comply with the WFD, which also had 2015 as its target date, specifically for the achievement of 'good status' and 'no deterioration'. In the NGOs' view, the 'old-fashioned' approach favoured would guarantee that in many cases the WFD's goals would not be achieved, 'while also failing to achieve truly sustainable flood management'. The NGOs went on to criticise loopholes, such as allowing member states the possibility of delaying the finalisation of management plans until 2021. The risk assessment procedures laid down in the Flood Directive were also inadequate, because they failed to give due consideration to the role of flood plains, to climate change and its potential impact, or to what they describe as the 'environmental and resource costs of technical measures' such as those arising from 'loss of drinking water sources, biodiversity, (or) recreation facilities'. Risk mapping would not be obliged to examine 'all relevant scenarios of flood events'. There was no obligation, either, for management plans 'to prioritise WFD objectives or measures for natural retention', meaning that it would be 'too easy to continue with old-fashioned technical measures'.

The EEB and the other NGOs predicted that the new directive would be a failure, allowing the risk of damaging floods to continue to escalate in all European river basins, 'including the Danube, Ebro, Elbe, Meuse, Oder, Rhine and Thames'.[42] This is especially the case as current climate change scenarios predict that extreme precipitation events, as we saw in Chapter 4, will increase in frequency and intensity. Yet the Commission was well aware of this and had been repeatedly warned of the impact of climate change. As one expert witness at a Commission workshop in 2006 fairly typically observed:

> Especially the alteration of flood frequencies and intensities could lead to a number of changes in flood risk systems. They may require a rework of the hydrological, statistical and hydraulic baselines for the determination of flood hazards and design levels. If water levels increase, enhanced efforts to maintain the current design standards of flood prevention would be needed. For example, this could lead to an enlargement of active floodplains, the re-design of reservoirs and flood polders, heightening and strengthening of flood defence structures etc. Moreover, secondary effects could be expected like changing benefit-cost-ratios for measures and instruments of risk reduction for both resistance and resilience strategies.[43]

Given the amount of rhetorical attention the European Commission has paid to climate change and related issues, it is perhaps surprising to note how little real impact it has had on policy. This is particularly the case when one considers that, according to a wide range of sources, the weather in Europe is becoming more extreme and less predictable. A majority of EU member states and applicant states were affected by the persistent floods of the 1990s and since, with some of the worst occurring in the 'old' member states of France, the UK and Germany, others in the Czech Republic and Hungary, and still more in the most recent entrants, Romania and Bulgaria. According to the Association of British Insurers, which has an obvious interest in the matter, the cost of flooding in Europe could increase to between €100 and €120 billion by 2080.[44]

The EEB believes that the Flood Directive will 'put implementation of the Water Framework Directive at risk, whereas more sustainable flood-management would support achievement of the directive's objectives'. They point out that projections indicate that 'over 70% of Europe's surface waters will probably not achieve "good" status or will not be saved from further deterioration if additional measures are not introduced to better protect water'. The Flood Directive will make things worse, because the hydromorpholigical effects of 'technical flood-protection measures, are among the greatest threats to water bodies'. They accuse the majority of member states of planning such measures without having made any 'comprehensive assessment of environmental and resource costs'. Evidence from Germany and the Netherlands in particular indicates that technical measures such as polders, dikes and dams are both extremely expensive and ultimately ineffective. Among the results of this misguided approach, the EEB concludes, would be 'a strong risk of further decline in floodplains and natural water courses and their associated biodiversity, which is very high, often with over 10,000 species'. In addition, 'drinking water sources, scenic recreational areas and room for sustainable water and land-use' would all be put 'at risk'.[45]

Commenting on the new directive, Christian Schweer from Friends of the Earth Europe said:

> As climate change increases the risk of floods, and water and land keep being used with insufficient consideration for natural ecosystems, it is likely that severe floods will hit Europe more frequently. Sustainable use of the whole river basin – particularly preserving and restoring floodplains – is the only efficient way to manage flood risks, while building concrete barriers to constrain rivers is short-sighted and expensive.[46]

As Pieter de Pous told us, the problem with the Floods Directive is that it fails to exhibit any real long-term perspective. 'In long-term thinking, you know that we would need more space for the rivers…you can't continue to build higher and higher dikes without, at some point, reducing pressure and creating emergency flood zones.' Physical flood defences should only be built where they are environmentally sustainable and economically justified, and decisions as to whether they were either of these things should be open and transparent. 'When it comes to new developments in flood-prone areas where you are building more houses or having a number of farmers farm in a way that is not compatible with the fact that the land is being flooded once in a while, then we suggest that flood defence methods probably shouldn't be built there. It concerns how you would go about new flood defence infrastructure and the kind of decision-making that takes place before you go in that direction.'

As we saw in Chapter 2, the best defence against flood is almost always to ensure that natural river basin functions are preserved or restored. By restoring natural forest cover and wetlands, protecting soil quality and encouraging environmentally-friendly management of agricultural land, flood defence can be incorporated into an overall environmental defence programme.

What Should be Done?

As the environmentalist NGOs argue, the Water Framework Directive, adopted at the turn of the century when 'green' approaches to policy problems were enjoying a vogue which touched even the political elites, does offer the possibility of improvement in the management of water in Europe whether from a social, economic or environmental point of view. Clearly, since then, two contradictory trends have combined to threaten the directive's potential as a sound basis for effective measures. On the one hand, under pressure from corporate industry and agriculture, member states have been retreating from commitments made when the WFD was adopted. This has been enhanced by the admission of new member states with weak environmental policy frameworks and outmoded water supply and sanitation systems.

On the other hand is accelerating climate change, which calls into question the adequacy of what was always a compromise text.

The first priority must be to ensure that everyone within the European Union has reliable access to clean, wholesome, affordable water. How this water is paid for must be left to each individual member state to decide according to its own traditions, but it should embody the principles of 'the polluter pays', 'the beneficiary pays' and 'to each according to need, from each according to ability to pay'.

River Basin Management Plans should be designed primarily to reduce demand, a goal which privatisation makes much more difficult to achieve. Companies operating for profit have a clear interest in *increasing* demand and none at all in reducing it. Reducing water demand through increasing the efficiency of delivery systems and of the way it is used, however, has other enemies in the corporate world. It is extremely difficult to make money by persuading people to use less of something. Encouraging increased demand is better from a corporate viewpoint, especially when this involves the need for dams, reservoirs, water relocation projects and other major engineering works, all of which must be built by private corporations at the public expense. Dams, like privatisation, are excellent devices for transferring public wealth into private pockets.

Beyond conservation and efficiency measures, the protection of supply sources – groundwater aquifers and surface water from rivers and lakes – from pollution and overuse must be urgent priorities. To minimise the impact on the public purse, the major polluters, which in Britain include privatised water companies, must be made to pay.

Farmers' dependence on agricultural subsidies could also be used to encourage better water management. Cross-compliance measures, which allow farmers to receive direct payments in return for meeting environmental, food safety and animal health and welfare standards, could be extended to include water conservation. Unfortunately the Commission rejected this approach following opposition from the farm lobby and those member states where it is influential.[47]

Unless a way can be found to implement such measures in the face of the power of the farm lobby, it is difficult to see how progress can be made and sustained. In addition to the restructuring of subsidies, the price mechanism must be used to encourage efficient use in agriculture, industry and energy production, with penalties for profligacy and waste. Householders should certainly be encouraged to be conscious of the water they use, but not told that they are a major part of the problem. With sensible and fair conservation measures, governments can take the lead in this, but raising people's awareness of what they can do as individuals should be accompanied by honesty about the real culprits. Financial assistance could be offered to households to encourage them to invest in water-saving devices, but industry and agriculture should be forced to adopt the most water-efficient technologies and binding minimum efficiency standards set for water-consuming equipment. EU structural fund money, currently made available for projects which seem to stand in open contradiction of the Water Framework Directive's requirements and goals, should instead be spent on developing comprehensive programmes which see the protection of Europe's waters and the systems which deliver water to where it is needed as the cornerstone of all progressive politics. Not only the defence of the environment, but social justice, economic equity and every human being's right to a dignified existence depend on the adoption of such a perspective.

7

THE EUROPEAN UNION BEYOND ITS BORDERS

'In the GATS negotiations on further liberalisation of services, currently underway in the World Trade Organisation (WTO), the European Union has asked 72 WTO member states to open up water delivery and waste water management to international competition. Liberalisation of water markets through the GATS talks would not only help Europe-based water TNCs to expand further. Bringing water supply under WTO disciplines may effectively make privatisation irreversible and close off the development of participatory and cooperative models. The EU's aggressive promotion of the narrow commercial interests of EU-based water corporations, spells disaster for the worlds poorest.'

Corporate Europe Observatory[1]

For the European Union to make a real impact on the water situation in developing countries, it would have to reform its policies across the board. As is the case within the EU, the availability and quality of water are affected by decisions taken in relation to transport, agriculture, industry, trade and a host of other areas. The EU's decision to impose a minimum quota on the use of bio-fuels, for example, will have disastrous consequences for rural communities and the environment in many commodity-producing nations. Space unfortunately means that we cannot go into these matters in detail, but it is important to bear them in mind when considering those policies influencing more directly the supply of water and sanitation.

Nobody should lack access to fresh water and decent sanitation, nobody in the world. There is simply no point in devoting resources to any other development goal until this fundamental human right has been achieved. Failure to achieve it is not an accident and it is not a result of human error or incompetence. It is the result, purely and simply, of a system designed to line the pockets of the few at the expense of the many. Every child, woman or man who must go thirsty, everyone who lacks a decent place to bathe or an efficient way to carry away their waste, everyone who must go hungry for lack of water to grow or cook their food, each is a victim of acts of violence, just as surely as those who wake up one day to find bombs raining down from their skies. And it is often the same people to whom these terrible things happen and the same people perpetrating the crimes, for essentially the same ends.

These acts of violence are committed by men and women, often in the name of institutions which claim to be trying to help. They may be private corporations pursuing profits on the back of the most basic human needs. They may be individuals in the pay of those corporations. They may be public bodies aiding these corporations, instead of using the power and wealth at their disposal to address this urgent crisis. And one of these public bodies is the European Union, whose unelected executive, the Commission of the European Communities, consistently puts the interests of European transnational corporations before those of solidarity or compassion.

As we have seen, water supplies and sanitation in almost every EU member state remain either exclusively or predominantly in public ownership. Resistance to privatisation in the face of European Union enthusiasm for the private sector has been marked, and largely successful. In France, earlier privatisations are even being reversed.

Within the EU, the European Commission continues to insist that it is neutral on the question of public or private ownership. This is disingenuous, because the entire economic ethos of the European Union is based on the idea that 'competition' is the key to economic success. Founding an economy on this principle, and attempting to introduce it artificially in the face of a natural monopoly such as water, involves getting rid of most of the advantages of public ownership. As water campaigner Olivier Hoedeman of Corporate Europe Observatory (CEO) told us, 'The EU and World Bank want to force public companies to behave as if they were private companies, to outsource whatever can be outsourced, to get rid of a real public element by either privatising or transforming the public companies. They want to see public companies commercialised.'[2]

Outside its own borders the European Commission seems to have no real interests other than boosting the market for European goods, European services and European corporations. European public utilities are on average much more efficient than their private equivalents, as we have seen looking at the UK and France. It is, however, the massive transnational corporations, such as Veolia and GDF Suez, which have the biggest armies of lobbyists in Brussels, the most to spend on propaganda and the means to buy the support and loyalty of legislators and officials.

When we interviewed Olivier Hoedeman at his office in Amsterdam, he told us that at the beginning of the century, when CEO first turned its attention to water issues, 'the Commission was liaising with large water companies, such as Vivendi [now Veolia] and Suez [now GDF Suez]. They entered intense dialogue in the year or so before [the Earth Summit in 2002] in Johannesburg, inviting these companies to help them shape the European Water Initiative. Commissioner Wallström was involved in this. They believed that the private companies were crucial, so they allowed them to help determine the way in which the EWI would be, how it would be carried out. This is just what happened between 2001–2002.'[3]

One means whereby the Commission attempts to further the interests of European corporations is to bring pressure to bear for inclusion of the water sector within the scope of the rules of the World Trade Organisation (WTO). Despite insisting on the exclusion of its own water sector from negotiations on the General Agreement on Trade in Services (GATS), the EU's negotiators have ruthlessly pursued developing countries, attempting to use the Union's economic muscle to bully these countries into allowing European TNCs access.

The problem, as campaigners like Hoedeman see it, is that 'the global private water market is entirely dominated by European water giants and the European Commission sees it as its task to assist the further expansion of these corporations, apparently regardless of the impacts in developing countries'.[4] This bias is driven by the intensive lobbying of such corporations. Perhaps frustrated by their continued failure to make progress within Europe, the water privatisers have intensified their efforts in recent years, establishing the International Federation of Private Water Operators (also known as AquaFed), with an office directly opposite the main European Commission building in Brussels.

If you add together the aid budgets of the member states and that of the EU itself, you come out with the biggest donor of aid for water services in the world. Any generosity behind the some €1.4 billion earmarked for water services, however, is reserved for the corporations which are its real beneficiaries. In recent years, governments have dedicated much of this aid to promoting the expansion of the private sector, presenting it as Hoedeman puts it 'as the way forward for water delivery in developing countries'.

Hoedeman's organisation, which monitors the behaviour of European TNCs globally, has uncovered a number of examples which provide evidence for this. What makes him particularly indignant is the enthusiasm for privatisation in other parts of the world of governments such as that of the Netherlands, where the law rules out any role for private operators in domestic water supply. He detects what he calls 'a similar hypocrisy' in the case of Germany, citing the example of El Alto, Bolivia, where Suez had its water concession terminated following seven years in which the company had failed to fulfil agreements or deliver promised improvements. The local population wanted a public utility with citizens' participation, but German state aid agency GTZ made loans conditional on the continued participation of Suez in management.[5]

Hopes for a change of heart and policy flared briefly when Belgian Liberal Louis Michel became Development Commissioner. Michel already had the sympathy of sections of progressive opinion as a result of his outspoken stand, as Belgian Foreign Minister, against the war on Iraq and was quoted as saying that he was 'with those who don't think that everything should become a commodity, and these services should be exempt from market pressures'. In reality, nothing changed. The EU's external water funding programmes –

the European Water Initiative and European Union Water Facility – remain much more about aiding European TNCs than they are about solidarity with poor people or poor countries.[6] During the design of the EWI, the Commission, Hoedeman says, 'were asking Veolia and the others, "How should we do it?" So the EWI shows the same ideological assumptions.'[7]

This has been graphically demonstrated by the failure to support calls for access to water to be declared a human right. For example when, at the Fourth World Water Forum in Mexico City in March 2006, Bolivia proposed that water be declared 'a fundamental human right' and that it should be excluded from trade agreements, France, the UK and the Netherlands joined the United States in vigorous opposition. The EU itself took the line that water, though 'a primary human need' was not a human right. This was in defiance not only of common sense and common humanity, but of a European Parliament resolution calling for recognition of just such a right.[8]

Yet as Hoedeman noted, if resolutions passed by the EU's only elected representative body have little impact, the same cannot be said of the abject failure of privatisation to deliver the goods. The Commission's beliefs 'that privatisation would deliver expertise and finance, that it is inherently more efficient' have 'proved to be false'. This has had, he said, some effect on Commission staff working directly on water issues, but others, particularly DG Trade, continue, in the face of overwhelming evidence, to adhere to their ideological commitment to privatisation. By 2005, Hoedeman told us, 'the dynamics had changed because of the failures'. A vigorous campaign to defend public ownership had also had its effects, but 'the fundamentals had not changed'.[9]

One problem has been that private corporations generally have not been able to raise finance. 'They can invariably not borrow as cheaply, they have to pay more for finance than governments do,' Hoedeman pointed out. High finance's recent severe crisis can only exacerbate this problem. In any case, the instruments used to fund investments in a sector such as water must be carefully chosen. 'Private equity is particularly inappropriate in a sector that requires a long-term commitment,' Hoedeman argues.

Public resistance to privatisation has been even greater outside the EU than it has within. As David Hall and PSIRU colleagues noted in an article for the journal *Development in Practice*, 'The collective political impact of the campaigns against privatisation is remarkable.' In South Africa, a water contract 'was nullified as public or municipal consent was never obtained'. In Paraguay, the national parliament had voted overwhelmingly 'to suspend indefinitely the privatisation plans for the state-owned water company'. In Brazil, the National Front for Environmental Sanitation (FNSA) brings together trade unions, public sector managers, professional associations, NGOs, consumer groups and other social movements in a campaign committed to keeping water supply and sanitation in the public sector.[10] In total, the PSIRU researchers list 29 successful campaigns against privatisation, 17 in developing countries, including nine in Latin America. Eight campaigns were in EU countries – both 'old' and 'new' member states – and the rest in North America. The campaigns involved, in a range of combinations, much the same elements as those participating in Brazil's FNSA: trade unions, consumer and citizens' groups, environmentalists, progressive political parties, organisations representing small business people and various NGOs with specific concerns over the likely impact of privatisation.[11]

Despite all of this, Hoedeman has retreated from his earlier belief that the campaign against water privatisation had been an almost unqualified success. When the latest round of GATS negotiations got under way, the Commission simply asked the TNCs for their wish-list. 'They sent them a questionnaire, and the questions covered such matters as access to groundwater, so they were interested in resource use, not only in delivery. The Commission made it clear that they would be willing to offer support if there was hindrance when it came to access to water. They followed-up with the questionnaire and asked for clarification; they took the process very seriously.' He now calls his earlier claims of success 'a mistake', adding that whilst 'it's true that the anti-privatisation campaign had had its influence' he had been 'too optimistic'. Some people went even further, switching their attention to other issues.

So why did the EU's attempts to use GATS to enforce water privatisation fail? To understand this you need to know how the negotiations were organised. Countries, or blocs, would make 'offers' and 'requests'. Their offers must revolve around reduced tariffs and other liberalisation measures, as do their requests. As Hoedeman explained, 'A problem arose for them when they could not offer to open up the EU water sector. This was due to trade union pressure. So their offers fell short.' In 2006, they withdrew their liberalisation request.

This has merely left matters in a state of uncertainty, however, an uncertainty compounded by the utterly undemocratic way in which decision-making in European trade policy is conducted. 27 countries must first agree on a mandate, which is then supposed to guide the Commission's hand in its negotiations. This gives the unelected Commission huge power. 'The trouble is that it's very difficult to change the EU mandate,' Hoedeman told us. 'It's impossible, even.' He gave the example of Belgium, the country which had the strongest objection to water liberalisation. Despite this, the Belgians 'would not go so far as to insist on revoking the mandate'. Things did not improve when the failure of the Doha round of trade talks to establish a comprehensive multilateral agreement under the WTO drove the EU to seek bilateral agreements with each of its trading partners individually, or in relatively small regional blocs. Transparency was dumped completely. According to Hoedeman, the talks on the proposed bilateral Economic Partnership Agreements (EPAs) 'have gone completely underground'.[12] It was already clear, however, that the European Commission would be seeking to use these talks to force developing countries to open their water markets to European TNCs. There is no reason to suppose that the Doha Round's failure would change the Commission's stance and nothing which has happened since could lead us to believe that this has happened.

What is clear is that the Commission retains a blind commitment to privatisation as a 'solution'. They are, Hoedeman said, 'not drawing the right lessons, not willing to admit that they were wrong. They are still trying to launch Public–Private Partnerships, and looking for new ways to promote privatisation, for example through management contracts with the World Bank, which requires no investment whatsoever from the company signing the contract. These are franchise contracts. They are eager to develop a model, to make privatisation seem attractive. They are not willing to consider the option that the public sector can be relaunched, given the chance to redefine itself.'[13]

The Ideological Front

As Olivier Hoedeman and others were careful to note when we interviewed them, the European Union is not alone in its pursuit of an agenda centred on enabling corporations to gain control of the world's water services. An ideological front has been constructed consisting of the World Bank,[14] its lending arm the International Finance Corporation, its subsidiary regional banks, the International Monetary Fund, the World Trade Organisation and most western governments, not to mention the corporations themselves. The international financial institutions (IFIs) attach structural conditionalities to loans in order to ensure privatisation of water and other utilities. Western governments, including those of the EU, have attempted to use the WTO's General Agreement on Trade in Services (GATS) to force developing countries to open their service industries to western TNCs. Having failed for the time being in this attempt, the EU is now using negotiations on bilateral trade agreements to pursue the same ends. The EU's policies in relation to its poorer neighbours show a similar zeal for privatisation, again going beyond anything which has been agreed at WTO level, where, as a PSIRU report on the EU's 'Neighbourhood Policy' reminds us, 'government procurement is exempted from the scope of the agreements on trade in both goods and services, and planned discussions on a multilateral agreement on government procurement have not yet started'.[15] In the Neighbourhood Policy, however, the European Commission takes full advantage of the fact – at least according to its own interpretation – that it is freed from the constraints of an EU Treaty which obliges it not to take sides on the question of public versus private ownership. As PSIRU notes, under the Neighbourhood Policy 'the action plans go beyond the limits of the EU treaty, and include a number of references to encouraging privatization as an objective in neighbourhood countries, both in general and in relation to specific sectors'.[16] Clearly, in the EU's worldview, there is a surprising amount of money to be made from the basic needs of the poor.

As the UN's Human Development Report 2006 pointed out, 'in countries with high levels of poverty among unserved [i.e. without a water supply] populations, public finance is a requirement for extended access regardless of whether the provider is public or private'.[17] This, then, is the key to understanding how private corporations hope to 'profit' from supplying the poor with water. The money will come out of the pockets of western taxpayers in the form of development aid. Instead of going to the people development aid is allegedly intended to help, large slices of this money will go to the shareholders of transnational corporations. This is scarcely 'profit' by the normal definition. It is, rather, a handy way of transferring public money into private pockets, of filling the corporate trough at the public expense.

That the drive to privatisation is led essentially by an ideological impulse can again be demonstrated empirically, just as was the case within the EU, by looking at its actual record. We have already seen the disastrous consequences of ideologically-driven water privatisation in the UK and the way the French system, by offering contracts to the private sector, fuels corruption and fraud. In 2008, according to PSIRU's Emanuele Lobina, there was 'an explosion of European cases where private water companies (were) being investigated... for corruption or serious irregularities'. Lobina mentions Turkey and Armenia, but also core EU member states Italy and the UK. In developing countries, however, the relative weakness of regulatory authorities in the face of the power of foreign corporations has compounded the problems associated with their involvement in water supply.[18]

A 1999 study by PSIRU found that while 'private sector involvement in water supply and sewerage is expected to facilitate investment and enhance the efficiency of operations' it has in practice 'exhibited economic problems and distortions characteristic of monopolies' including 'management inefficiencies, restricted competition and corruption, excess pricing and restricted access, excess profits and low water quality, and problems in delivering development objectives'.

Examples given include:

- Trinidad and Tobago, where delegation of the management of the islands' water to a subsidiary of British water corporation Severn Trent was brought to an end after five years in which time there had been 'no significant improvement in the reliability of supply and coverage of sewerage';
- in Puerto Rico, a contract to manage the water and sewerage authority was given to a subsidiary of Générale des Eaux (now Veolia). The contract was criticised in a report issued by the Puerto Rico Office of the Comptroller in August 1999, for 'numerous faults, including deficiencies in the maintenance, repair, administration and operation of aqueducts and sewers', while the authority's operating deficit had rocketed 'without any noticeable improvement in the service';
- in Côte d'Ivoire, a subsidiary of another French water corporation, SAUR/Bouygues, was in 1987 given the contract to supply the entire country with water, no competitive tendering process having taken place;
- in the Philippines, on taking over water management, a subsidiary of British corporation Biwater increased water rates to industrial customers by 400 per cent;
- in the Argentinian province of Tucuman, when another subsidiary of Générale des Eaux was granted a 30-year concession to supply the region, water tariffs doubled. Despite this 'the company failed to accomplish the planned investment programme allowing the water supplied to turn brown'.[19]

A separate PSIRU publication, *Water and Privatisation in Latin America, 1999,* listed nine examples of failed privatisations in Latin America, each resulting in either a terminated or renegotiated contract, or the abandonment of the planned transfer to the private sector. In the latter case, plans were changed following widespread opposition. In the former, problems included poor performance, high costs and failure to meet connection targets.[20]

A report prepared by US group Public Citizen investigated cases in which 'showcase water privatisations have suffered major losses' and found examples of what it described as 'fiascos' in Buenos Aires, Manila, Atlanta and Cochabamba. As the report 'Water Privatization Fiascos' concludes, '(it) has now become clear...that the major multinational water corporations have no intention of making a significant contribution to the capital needed to ensure access to clean and affordable water. The rhetoric of private sector financing is a myth.'[21]

A study of water privatisations in Africa demonstrated that the main effects were increased prices and disconnections, concluding that these 'must mean that the poorest segments of society are likely to be the main losers from the privatisation process'. The researcher pointed to weak regulation, in particular the absence of an independent regulator, the failure to clearly delineate and define responsibilities and the broader context of a weak institutional framework, as guaranteeing that privatisation would not achieve its stated objective.[22] The cases of France and the UK, however, show that even in highly developed societies similar problems can arise.

One reason for these recurrent problems is the simple fact, to which perhaps we should apologise for returning so persistently, that in the water sector privatisation does not bring competition. As a PSIRU study from 2000 concluded, it brings not competition but 'collusion between a very small number of companies – worldwide...'. The world market is, as is the case with the privatised water sector in the EU,[23] dominated by a very small number of TNCs. Worse still, they often operate jointly and give any outsourcing contracts to their own subsidiaries. The PSIRU report goes on to list ten documented examples of such collusion, most of them within the EU, the basic reason for this being that 'the dominance of Vivendi and Suez-Lyonnaise is such that even the world's largest multinationals find it impossible to enter the market – except in partnership with the French companies'.[24]

Corruption, as we saw in Chapter 5, is the almost inevitable accompaniment to the award of the long-duration contracts made necessary by the nature of the sector. 'This may take the form of a bribe to a person or a political party, or the allocation of monopoly profits to politicians or their relatives.' The weaker the regulatory structure, the more vulnerable is the process to corruption of this kind. But as PSIRU notes, it can happen anywhere, and where it happens it can quickly become structured into the system.

On price, while 'this is usually the key parameter used to evaluate bids for private water concessions...after the bidding process, the operator has a simple incentive to maximize prices'. Bargaining with the regulator, or conducting 'technical adjustments and renegotiations which cumulate to a considerable gain' are common methods of achieving this. Errors in forecasting are often blamed, even where these are clearly the company's own fault. For example in Dolphin Coast in South Africa, French transnational SAUR 'informed the authority that the failure of middle class development to keep pace with their assumptions means that prices have to rise'.[25] Once a relatively underdeveloped country has opted to hand its water supply to a foreign corporation, however, the latter is clearly in a very powerful position.

This leaves governments in a weak position when contractual obligations are not met. PSIRU gives a number of examples of the effects of this:

> In developing countries the key public service objective is usually to extend the service. Recent studies in Cartagena (Colombia), Cordoba (Argentina) and La Paz (Bolivia) have all shown how private operators (Suez-Lyonnaise in all cases, as it happens) systematically avoid making the investment in such extensions, despite contractual obligations and political demands, because they regard such connections to the urban poor who can only afford water with cross-subsidy as too risky – not profitably sustainable.

Thus the chain of responsibility is simply broken by the force majeure of profitability.[26]

This position is backed up by pressure from aid donors to privatise their water supplies and other water services. Because of increasing reluctance by TNCs to invest in water in developing countries, these countries have responded by allowing corporations to cherry-pick those parts of the system which they can run at a profit. This usually means a management contract, in which the foreign corporation manages water supply, including metering, billing and payment collection, while the state continues to take responsibility for infrastructure. Such contracts operate in South Africa, Mozambique and a number of west African countries.[27]

Amongst these donors is the European Union. Based on a plan originally devised by a panel chaired by one-time IMF director Michel Camdessus, the central feature of the EU's plans for its development aid to the water sector was the diversion of moneys to promote the entry of European private water corporations into markets in developing countries and transitional economies outside the EU. Written by James Winpenny, an independent economic consultant who has written reports for the OECD, the Camdessus Report was presented to the 2003 World Water Forum in Kyoto and met with a mixed response. Despite this, it provided the inspiration for the EU's drive to privatise the world's water and see it controlled by European TNCs. The report noted that the Camdessus panel was 'convinced of the vital importance of private sector disciplines, know-how and management skills in the reform and further development of the water sector', though it did add that each case should be taken on its merits.[28] Even the development minister of the right-wing Dutch government, Agnes Van Ardenne, described the report as 'disappointing', while development economist Sir Richard Jolly, chair of the Water Supply and Sanitation Collaborative Council (WSSCC),[29] said that it contained 'too little on what is needed to reach the poor and the poorest of the poor'.[30]

Shortly after the Kyoto World Water Forum the European Commission released proposals for the establishment of the

European Water Fund. It was initially to be worth €1 billion and would be used to finance investments in the ACP countries, the 77 developing countries whose 'special relationship' with the EU derives from their former status as colonies of one or another of what are now its member states. The proposal envisaged using a mix of 'guarantees, risk insurance, soft loans, etc' as a means of facilitating the intervention of EU operators and sought to build on the EU Water Initiative (EUWI) launched at the UN's Johannesburg Rio+10 Summit the previous year.[31] This 'Initiative' was designed to rechannel over €1.4 billion from EU development aid funds into the financing of water supply Public–Private Partnerships (PPPs) in Africa, Latin America and the former Soviet Union. One member of the EUWI finance panel was James Winpenny, author of the Camdessus Report, while Alan Hall of the private sector Global Water Partnership was coordinator of its financing working group and was also one of the authors of the corporate-controlled World Water Council's paper on governance and water, presented at Kyoto.[32]

Behind the scenes, the Commission was developing these initiatives in close collaboration with transnational corporations in the water sector. 'Confidential documents' obtained under a freedom of information procedure by the Corporate Europe Observatory 'revealed how the European Commission has from the early stages had close consultations with major water TNCs about the EUWI'. These corporations, which included Suez, Thames Water and Vivendi (now Veolia) had been 'deeply engaged in discussions on the EUWI with Commission representatives at all levels' participating in most of the EUWI working groups. Suez was particularly active, which may have been helped by the fact that Ives Thibault de Silguy, the Suez board member liaising with the Commission, is very much at home in Brussels: from 1994 to 1999, he was himself a European Commissioner. The Commission also involved Suez and other water sector TNCs in preparing its negotiating position for the WTO Doha Round,[33] stating at the time that 'the main objective...for the negotiations is to reduce the barriers which European operators face in third countries' markets'.[34]

The European Commission lists the EUWI's five objectives as follows:

(i) the reinforcement of political commitment towards action and innovation oriented partnership; (ii) the promotion of improved water governance, capacity building and awareness; (iii) improved efficiency and effectiveness of water management through multi-stakeholder dialogue and coordination; (iv) strengthened co-operation through promoting river basin approaches in national and transboundary waters; and (v) identification of additional financial resources and mechanisms to ensure sustainable financing.[35]

The essentially bureaucratic nature of these goals means that whether or not they have been achieved is a secondary question at best. If their achievement fails to provide the tools for getting more and higher quality water to more people in poor countries and poor communities the question is rendered irrelevant. And the fact is that the statement in a 2005 joint report by Water Aid and Tearfund that 'not a single extra person has received safe water or sanitation through the initiative' remains true.[36] This, and not the bureaucratic 'progress' trumpeted in EUWI's Annual Report, is the true measure of the success or otherwise of the initiative.

One of the reasons for the EUWI's lack of success is that it relies on the EU's member states entering into partnership dialogues with developing countries and few have taken up this challenge in any sustained way. Another is the virtual exclusion from such dialogues of civil society organisations from those countries which are supposed to benefit from the initiative. The most important reason for the failure, however, is what the Tearfund/Water Aid Report refers to as the 'ideogical bias to private financing'. Although the Initiative was ostensibly established to improve the use of the EU's own €1.4 billion of annual aid for water-related development projects, its priority has always been to attract private finance. The problem is, as the report – and many others – have pointed out, 'there is no evidence of international investor willingness to finance developing country water projects'.[37]

The EUWI is not simply hampered by the reluctance of the private sector to commit itself financially, however, or even by its own bias towards TNCs as a source of finance. It was from the start ill-conceived, its bureaucratic structure and lack of focus hampering any possibility of progress. In 2004, in an attempt to overcome the funding problem, the EU established a separate European Union Water Facility (EUWF), receiving €500 million from the member states.[38] The reasoning behind this was an admission by the Commission that 'progress made is not sufficient to meet the Millennium Development Goals without a significant increase in financing and improved mechanisms to help development aid attract other resources (private sector, development banks)' and that 'the creation of a European Water Facility will serve both these purposes'.[39] In reality, while the EUWF has been more successful than the farcical 'Initiative', it has relied almost entirely on public sources of money. As a mid-term evaluation report concluded:

> The scarce participation of the private sector, *which was expected for the water sector*, is due to well-known management and structural difficulties, and deserves special attention. The use of a different definition for private sector could be an option to facilitate the participation of communities, as 'private investors' as so is the opportunity for other small-scale financial forms. (Authors' emphasis)[40]

The pattern is clear. Expert advice tells the Commission that the private sector will not come up with the goods, having had its hands burnt rather too often. The Commission, however, attempts to make life as easy as possible for its friends and former colleagues in the European water TNCs. Their limited response to this is inadequate to the task. Though they may continue to look for opportunities in such places as North Africa and the ex-Soviet republics, they have no interest in the very poor countries which are precisely those upon which development aid for water and sanitation should be focusing.

What is obvious is that only solid commitment from the EU's member states to invest in the achievement of quantifiable targets, mobilising where appropriate the expertise and resources of their own publicly-owned water companies, can hope to make progress towards the Millennium Development Goals. Yet, like the EUWI, the EUWF lays emphasis on the continuing, futile search for private investors.

Public–Private Infrastructure Advisory Facility (PPIAF)

The EU and many of its member states are also among the biggest enthusiasts for World Bank funding mechanisms linked to privatisation conditionalities, such as the Public–Private Infrastructure Advisory Facility (PPIAF). Established in 1999 by the World Bank, the United Kingdom and Japan, the PPIAF has been used as the instrument for funding water privatisation processes in 37 countries in the developing world. The UK remains by far the biggest donor, but the Facility has also received moneys from the European Union and from other EU member states: Sweden, the Netherlands, Germany, France and Italy. PPIAF ignores possible public sector solutions and concentrates on advising developing countries as to how they can privatise their water supply and, hopefully, attract foreign direct investment to that end.[41] In this it flies increasingly in the face of evidence that this approach simply will not deliver.

In a report for the World Development Movement published in 2006, David Hall and his PSIRU colleague, Emanuele Lobina, explained just how little success the favouring of the private sector could claim. In order to understand this, it is first necessary to appreciate that 'privatisation' is a broad term which covers a multitude of sins. The privatisation of water supply or sanitation services generally takes the form of a contract between public authorities, which may in some cases be the existing supplier, and a private company or companies. It is difficult to find examples of the wholesale privatisation of water supply and sanitation, where a company purchases the actual infrastructure, outside of England and Wales. The sale of the water itself, including lakes and rivers, as proposed by Turkey's government, is unprecedented.[42] The norm is the contract between a responsible public body and a private service provider. These contracts vary greatly in the obligations placed on the private supplier, but fall into three categories.

In the first, the 'concession contract', a company is given a licence to run the water system and takes responsibility for any necessary investment. It may be required to ensure that more households or other kinds of customer are connected, that connections are improved, that the quality of the water supplied is improved and so on. Such contracts typically last 25 or 30 years, but may be considerably longer.

In the second, the 'lease' or '*affermage* contract', the company takes responsibility for the water distribution system, for investment for repairs and renewal of existing assets, but not for new infrastructure, which remains the responsibility of the public authority. Any new household connections, therefore, lie outside the scope of what it is expected the private company will finance.

Finally, the 'management contract' leaves all investment to the public authority, with the private firm taking responsibility only for managing the system, or aspects of it. As Hall and Lobina say, such contracts 'are risk-free for the private sector and do not involve any investment by the private company'.[43]

These technical details are important because it is the claim of each of the participants in the PPIAF, including the EU and its member states, that their reason for participating is to achieve the Millennium Development Goal on water and assist in the achievement of others. Yet the MDG speaks most importantly of reducing the numbers of people without a reliable fresh water supply by half by 2015. Only one of these three kinds of contract will help to achieve this, though the second may help to curb any local declines in connection rates. Essentially, however, 'only concession contracts bring private investment to extend the system'.[44]

After reviewing investment in the sector in sub-Saharan Africa, South Asia and East Asia – regions in which 80 per cent of unconnected households are located – Hall and Lobina concluded that '(m)ost private contracts, notably lease and management contracts, involve no investment by the private company in extensions to unconnected households'. In one region, South Asia, not a single additional household has been connected to the water supply as a result of private sector investment. Even under concession contracts, initial investment commitments are 'invariably revised, abandoned or missed'. Yet again we encounter the system of 'private gain, public pain', where public finance or financial guarantees are used to eliminate any real risk to the private corporation. In sub-Saharan Africa, four out of five contracts have been terminated or are mired in dispute over investment levels. 'Private water companies', the PSIRU researchers have established, 'do not bring new sources and volumes of investment finance – they rely heavily on the same sources as are available to the public sector'. As they say:

> This evidence debunks one of the most important myths concerning water privatisation, namely that private finance will play an important role in delivering progress towards the water and sanitation MDG. On the contrary, it has not done so up to now, and is unlikely to do so in the future.[45]

Worse, the effects of favouring the private sector have been negative in their 'impact on progress towards the water and sanitation MDG with major implications for communities of poor people around the world'. In the decade following 1997, only 600,000 households in these three critical regions were connected as a result of investment by private sector operators: an estimated 900 people a day. 'This is in contrast with the 1.3 billion people in those regions who it is estimated need a connection to a clean water supply between 2006–2015 in order to meet the water MDG.'[46] Despite this failure, the emphasis on the private sector, and the broader context of IMF-induced public spending cuts out of which it has grown, have led to what Hall and Lobina call 'a massive reduction in the level of aid and development finance from donors to the water sector which has far outweighed the actual investments made by private companies'. The result is that the 'net contribution of 15 years of privatisation has...been to significantly reduce the funds available to poor countries for investment in water'. The profit motive has also meant that even those connections which have been made have been the least costly available. The poorest countries, the poorest cities and the poorest neighbourhoods have been shunned. Sub-Saharan

Africa and South Asia have together received 'only one per cent of total promised private sector water investment'. The need to turn a profit has also meant that for poor customers connection is far from the end of the story. If they can't pay, they go back to being thirsty as pre-pay meters and disconnections deprive people of their supply. And as privatisation is almost invariably accompanied by massive price increases, failure to pay is the common consequence.[47]

The European Commission did not begin funding the PPIAF until 2005, bucking a trend which had seen a number of donors or potential donors having second thoughts. Its involvement is also in contradiction to its stated practice and philosophy which is laid out in a set of recommendations for reforming state enterprises published in 2003. This, in the Commission's own words,

> does not attempt to settle the debate on the advantages of different forms of ownership of enterprises – public, private or PPP. Rather it argues for the essential importance of looking objectively at all the options and their sequencing and selecting the one that best meets the needs of the particular country and field. This needs to pay careful attention to the capacity and resource constraints of the country.

It also notes the need to 'take into account the employment and social consequences'.[48] The PPIAF's privatisation-only policy is seriously at odds with the principles outlined in the Communication.

PPIAF has funded water privatisation processes in at least 37 countries, most of them in the developing world. In many of these, consultants have been paid for 'consensus building' work, the 'consensus' in question being in favour of privatisation. As Corporate Europe Observatory (CEO) points out, this involves interfering 'in internal democratic debates and deliberation on the provision of essential public services'. CEO also reports that in 'at least 17 countries where PPIAF has funded activities in the water sector, water privatisation conditions had been attached to loans, debt relief or aid, by international financial institutions like the World Bank and the International Monetary Fund'.[49]

BizClim and the Private Sector Enabling Environment Facility

Farmed out to a private consultancy, WYG International, and presided over by the ACP Secretariat and the European Commission, the €20 million Private Sector Enabling Environment Facility (PSEEF) seeks to go beyond PPIAF's scope. Its aim is to boost the private sector in the ACP countries under the broader ACP Business Climate facility, known by the hideous acronym 'BizClim'. BizClim's website states that it is dedicated to:

> improving legislation, institutional set up and financial measures (the rules of the game) relating to the enabling environment of the private sector in ACP countries or regions and to the reform of SOEs [State Owned Enterprises] – and to do so by focusing on possible support to ACP governments or regional institutions.

According to critics, however, this advice invariably concerns contraction of the public sector in favour of private firms. 40 per cent of the total PSEEF budget was initially allocated to the reform of SOEs.[50] Nevertheless, BizClim seems to have met much the same problem as did PPIAF: the private sector is simply not interested in investing in the poorest countries or the poorest communities, even with the active financial encouragement of the European Union.

This failure should not disguise the fact that, as CEO says, 'the European Commission seems to treat services of general interests – such as water supply – in developing countries as if these were just like any other sector of the economy'. It could be argued that in one sense the Commission should be doing just that. If the Treaty obliges it to be neutral on public versus private ownership within the EU's borders, then it is surely morally obliged to be neutral beyond them. This would indeed imply using the same measuring rod in each situation, each sector and each country. As this would lead to different results in each case, however, this is clearly not what is happening. The bias towards the private sector is clear, even in the names employed for the different facilities and programmes, let alone their actual behaviour. As CEO notes, 'BizClim is intended to act in synergy with...the EU–Africa Partnership on Infrastructure, the EU Energy Facility and the EU Water Facility, all of which involve very large amounts of EU aid funds and have a clear mandate to promote private sector participation in services delivery'.[51] BizClim has also a record of working closely with the EU employers' body UNICE, their African counterparts Association Industrielle Africaine and the African Business Roundtable. BizClim played a significant role in the 'ACP Business and Investment Forum', preparing the forum's agenda, identifying speakers and providing the European and African businesses with information regarding the business climate. The Forum was advertised as an ideal opportunity for the private sector to speak to high-level leaders and officials, *de facto* providing the private sector with privileged access to government officials, compared to civil society stakeholders.[52]

Despite all evidence that essential services cannot be efficiently delivered by private corporations, BizClim and the PPIAF continue to use European taxpayers' money to finance 'consensus building' activities, in reality exercises in pro-privatisation propaganda. The effectiveness of this propaganda is enhanced by the fact that in many cases it takes place in circumstances in which the choices available to the government in question have been seriously restricted by conditionalities imposed by an IFI or donor country.[53] The purported availability of private finance is no doubt welcomed by governments who have been told, quite simply, that they will not be permitted to raise funds for public investment. Meanwhile, the problem does not go away. And at its root is the fact that the poor cannot afford to pay the full cost of the water they need, so that until we achieve a more equal world, means will have to be found to pay for it. Nothing in the experience of 30 years of neo-liberalism has offered the slightest evidence that subsidising private corporations is an efficient means of achieving this. What we are up against is not evidence but ideology, and this ideology is killing people.

Public Alternatives

Investment in water supply and sanitation should surely be the priority for development policy. The United Nations has produced a list of potential benefits of improved water management which demonstrates this. This goes beyond the provision of fresh, clean water to households and places where people gather, whose health benefits will be obvious to anyone, especially anyone who has ever been deprived of such for any length of time. They include other relatively obvious gains, for example for agricultural and horticultural production. Less obviously, improved water management can lower the incidence of disease, including by reducing mosquito habitats and also pollution. All of these have clear knock-on effects by improving human health and therefore productivity, school attendance and even, as a result of reducing the number of women and girls obliged to spend large parts of their day carrying water to their homes, making a contribution to achieving progress towards greater gender equality.[54]

If these gains are to be realised, the first priority must be a clear recognition that water is a unique public good. The European Parliament, which has recognised this in a resolution on the internal market,[55] must insist on it in its approach to trade and development. Beyond this, however, it must be defined as a human right and legally recognised as such in national and international codes. Again, the European Parliament has recognised this, but the EU refuses to follow suit.[56] Rather than making aid conditional on privatisation, developing countries should be encouraged and assisted to create and realise such a right. This right must be linked to concrete action and statistical targets to measure progress towards their achievement.

Through both general development programmes and the encouragement of PUPs, developed countries should be assisted in developing 'national plans for accelerating progress in water and sanitation, with ambitious targets backed by financing and clear strategies for overcoming inequalities' in line with the UN recommendation of 2006. Given that such progress depends on 'large upfront investments with long payback periods' this can clearly only be achieved through development aid aimed at the public sector.[57] The same UN report calculated that aid flows would need to 'roughly double' if the MDG target is to be achieved, and that water and sanitation would have to figure much more highly in the list of development priorities.[58]

There is no single, inflexible model for best practice in the delivery of water or the provision of sanitation. Circumstances differ, in terms of social, economic, cultural, political, environmental and geographic realities. The model must therefore be adapted to these realities, but in each case it will be a public ownership model with a high degree of popular participation and democratic control. Profiting from a good or service that is vital to life is inappropriate, because the profit system can only function where the ultimate threat is withdrawal of the good or service in question. Private corporations must prioritise the interests of their shareholders. These can be served only by maximising profit and where the consumers in question are too poor to pay, such profits can be garnered, if at all, only through public subsidy. Public money spent in this way can be spent much better in the direct provision of water and sanitation. If the state spending this money is itself poor, a large slice of it will be development moneys funded from the taxes paid by citizens of rich northern countries, such as the authors of this book. And we believe we can assert with confidence that our fellow taxpayers share our preference for our money to be spent in ways which benefit those whose lack of access to water and sanitation means that they live with the constant threat of thirst, hunger and disease, rather than those who live with the threat of a lower-than-average share dividend.

8

A BETTER WATER POLICY IS POSSIBLE

The European Parliament declares 'that water is a shared resource of mankind and that, as such, access to water constitutes a fundamental human right; calls for all necessary efforts to be made to guarantee access to water for the most deprived populations by 2015'; Point 1, European Parliament resolution on the Fourth World Water Forum in Mexico City, March 2006[1]

'Don't get mad, get organised.' Big Bill Haywood, US labour leader

The provision of fresh, clean water and sanitation is not only the world's most pressing problem. It is the problem which must be solved before any of the others are tackled. We could run through the eight Millennium Development Goals and explain why each of them depends on going well beyond halving the number of people who do not have access to a supply of fresh, clean water; but that would be to repeat ourselves and to risk patronising our readers. The MDGs are readily available and you can easily see for yourselves the truth of our arguments.

We believe we have done enough to establish that water is a human right and must be recognised as such, not merely in words but in urgent practice. Following Peter Gleick,[2] we use the term 'human right' not loosely but to refer to a right recognised under international law. This precise usage makes it clear that both the nation state and international organisations have a clear duty to ensure that the right is granted. A Millennium Development Goal which plans to leave half of those who were without water and sanitation in 2000 still without them in 2015 is in reality one which condones a grave abuse of the human rights of hundreds of millions of people.

The human right to water is guaranteed by the United Nations under the Universal Declaration of Human Rights. It ought to be enough to say that none of the rights guaranteed within that document, from the 'dignity' of Article 1 to the 'right freely to participate in the cultural life of the community, to enjoy the arts and to share in scientific advancement and its benefits' of Article 27, can be exercised in the absence of reliable water and sanitation. Nor has the 'duty to the community' imposed by Article 29 been fulfilled if these essentials are not made available to all. We can be more specific than this, however. Two Articles in particular establish a human right to water and sanitation.

Article 23 (3) gives every worker 'the right to just and favourable remuneration ensuring for himself and his[3] family an existence worthy of human dignity, and supplemented, if necessary, by other means of social protection'. If such an existence is unavailable, then this right cannot be exercised. And Article 25 (1) states that:

> Everyone has the right to a standard of living adequate for the health and well-being of himself and of his family, including food, clothing, housing and medical care and necessary social services, and the right to security in the event of unemployment, sickness, disability, widowhood, old age or other lack of livelihood in circumstances beyond his control.[4]

This is even clearer, as none of these things is remotely possible without water and it is impossible to imagine anyone enjoying any of these rights unless he or she at least has access to a reliable water supply no more than a short distance from home, and some form of hygienic waste disposal.

The first explicit, official definition of the right to water appears to have been at the UN Conference on Water held in Mar del Plata, Argentina in 1977.[5] As this conference agreed that national plans should be adopted aiming to provide safe drinking water and basic sanitation to all by 1990, it provides a measure of how far the relevant MDG represents a retreat from this position. The conference declaration stated that '...all peoples, whatever their stage of development and their social and economic conditions, have the right to have access to drinking water in quantities and of a quality equal to their basic needs'.[6] There is a clear if less explicit recognition of the right to water in the UN Declaration on the Right of Development of 1989, which enjoins states to 'undertake, at the national level, all necessary measures for the realization of the right to development and ...ensure, inter alia, equality of opportunity for all in their access to basic resources'.[7] The 1989 Convention on the Rights of the Child commits states to combating 'disease and malnutrition, including within the framework of primary health care, through, inter alia, the application of readily available technology and through the provision of adequate nutritious foods and clean drinking-water, taking into consideration the dangers and risks of environmental pollution'.[8]

The declaration of the 1992 Earth Summit in Rio de Janeiro goes into considerable detail about the way in which the (by then recognised) human right to water should be realised. It notes 'water is needed in all aspects of life' and so the 'general objective' of the declaration 'is to make certain that adequate supplies of water of good quality are maintained for the entire population of this planet'.[9]

The means to achieve this is through 'integrated water management' which, the declaration says 'is based on the perception of water as an integral part of the ecosystem, a natural resource and a social and economic good, whose quantity and quality determine the nature of its utilization'. Because of this, 'water resources have to be protected, taking into account the functioning of aquatic ecosystems and the perenniality of the resource, in order to satisfy and reconcile needs for water in human activities'. Most importantly from a human rights point of view is the statement that in 'developing and using water resources, priority has to be given to the satisfaction of basic needs' as well as to 'the safeguarding of ecosystems'. While 'water users should be charged appropriately', this should be done according to means and with the involvement of the community, according to four principal objectives: the promotion of 'a dynamic, interactive, iterative and multisectoral approach to water resources management', integrating 'technological, socio-economic, environmental and human health considerations'; 'sustainable and rational utilization, protection, conservation and management of water resources based on community needs and priorities within the framework of national economic development policy'; '[to] design, implement and evaluate projects and programmes that are both economically efficient and socially appropriate within clearly defined strategies, based on an approach of full public participation, including that of women, youth, indigenous people, local communities, in water management policy-making and decision-making'; and finally, the development of 'public participatory techniques and their imple-mentation in decision-making, particularly the enhancement of the role of women in water resources planning and management'.[10] Elsewhere, the declaration reiterates and adds to this, stating that 'emphasis has to be placed on the introduction of public participatory techniques, including enhancement of the role of women, youth, indigenous people and local communities'.[11]

What is clear is that in parts of the world where any progress whatsoever has been made towards the fulfilment of these objectives, this has been carried out under public ownership. The integrated planning which takes 'into account long-term planning needs' and 'incorporate(s) environmental, economic and social considerations based on the principle of sustainability' and 'constitute(s) an integral part of the socio-economic development planning process' is incompatible with the private ownership of water or of its means of distribution. The UNEP declaration underlines the fact that it is not solely that water is essential to life which makes its treatment as a commodity inappropriate, but the range of problems which derive from this indispensability. While UNEP wants to see sustainable management of water and recognition in all planning decisions of its full cost, including 'investment, environmental protection and operation costs', it also insists that in the design of a payment structure, 'charging need not necessarily burden all beneficiaries with the consequences of those considerations'. Charging should not only reflect 'the true cost of water when used as an economic good' but also 'the ability of the communities to pay'.[12]

Attempts to achieve these ends under a system which offers ownership and control of water resources to profit-seeking foreign corporations can only result in bureaucracy, inadequate supervision and 'public pain, private gain'. The private corporations themselves have increasingly realised that they can neither fulfil these objectives while making a profit, nor get away with failing to do so. The failure of privatisation of water supplies in developing countries should not have been necessary, as a study of the 1992 declaration would lead any honest person to conclude that private corporations were entirely inappropriate instruments. Now that we have seen so many failures, however, we are held back only by the systematic bias of international donors, amongst which the European Union is one of the worst culprits, in favour of the failed private model. A variety of innovative systems of public ownership have demonstrated that privatisation is not a necessary solution to the problems which have characterised top-down nationalisation models in developing countries in the past. The real solution is democratisation and popular involvement. The development solution is to remove pro-privatisation conditions from aid, cease to use trade negotiations to promote the interests of European TNCs and explore instead a mutually beneficial cooperation through PUPs, technology and know-how transfer and programmes based on ambitious, quantified targets which go beyond the current Millennium Development Goal. A world where no-one lacks clean water is within our reach.

Conclusion

For these reasons, the authors reject any role for private capital in the ownership or distribution of water or of water services of any kind, including sanitation. Water is far too important to be left to profit-making corporations. Most people in the world who can afford to pay the full cost of their water and sanitation, whether as individuals or communities, already have access. It is the poor who are missing out and lack of access to water is the most dramatic and dangerous demonstration of their poverty. A sound water supply and basic sanitation lead to immediate and dramatic drops in disease and their provision is the most straight-forward way to release the potential energy in individuals and communities hampered by poor nutrition and ill health. These essential services enable people to put an end to environmental degradation and raise themselves out of economic, social and environmental poverty. People are capable of solving their own problems if only they are given the space to do so. To provide water services to a community is to begin the rolling of a snowball of accumulating benefits which will repay investment again and again and again. Crucially, however, this repayment will not be to those making the initial investment, at least not directly. This should not matter to states, to regional or local public authorities or to providers of development aid. To private corporations, however, it is the beginning and end of what matters.

Privatisation is described by the Public Services International Research Unit as 'a failed experiment'.[13] Not wishing to imply any criticism of PSIRU, whose invaluable research has provided us with so much of the evidence we needed to reach and support our conclusions, we would nevertheless take issue with this description, which risks giving the impression that behind neo-liberalism lies a genuine, if misguided, attempt to address the problems of poverty and underdevelopment. In reality, neo-liberalism is merely the latest wave of theft which goes back at least as far as the enclosures which removed the English peasantry from their land and the Highland Clearances which achieved the same in even more brutal fashion in large areas of Scotland. The drive by the European Union, the World Bank and other international bodies to use their financial power to force developing countries to privatise their water services forms part of this shameful history. Their commitment to privatisation was never in reality about improving efficiency or getting water to more people. It has always been an ideological commitment, a set of ideas in the service of powerful interests, and a way of making money from people's basic needs.

A number of things went wrong. Private corporations had to be brought on board, and their interests are purely financial. When they discovered that they could not make money from privatisation, they hoped that it could be used instead as a conduit for the transfer of public money – in this case development funds – into private pockets. This came up not only against popular resistance and damning criticism, but against the hard reality that the private corporations simply could not deliver what they had promised. As we have seen, the rate of new household connections by private corporations is pitifully low and completely off target for meeting the MDG. In the 1980s, before the privatisation model became widely accepted, the public sector used the funding it received from International Financial Institutions and donor agencies to reduce the global proportion of people living without a safe supply of water from 56 per cent at the beginning of the decade to 31 per cent by its end. Expansion of water TNCs into developing countries' markets has now been reversed, with each attempting to unload water operations in the face of intractable reluctance from potential buyers.[14]

Throughout the world, disappointing returns on private investment have been compounded by the poor service delivered by the companies involved and by the popular discontent which has followed. In Bolivia, the Philippines, South Africa, several West African countries and Tanzania, privatisation has provoked popular resistance which eventually succeeded in seeing water supply returned to public ownership. In many other countries – Honduras, Nicaragua, Ecuador, India and Mexico, to name just a few examples – resistance is ongoing.[15]

Resistance is also growing in Europe, with the recent founding of a broad coalition against water privatisation. Founded at the European Social Forum in Malmö in September 2008, with an initial membership comprising groups from 17 different European countries, the European Public Water Network (EPWN) is working to transform EU policies towards water management.[16]

EPWN is merely the latest expression of the huge opposition which the privatisation of water is generating. This opposition is broad-based and many-faceted and it is forced constantly to align itself against international financial institutions, the OECD, the World Trade Organisation and the European Union, each of which consistently places the interests of powerful corporations before those of 'ordinary' men, women and children.

To place this movement in its historical context is to realise its significance. The usurpation of public property which began with enclosures is now carried out by transnational corporations and by governments acting on their behalf. Privatisation is just one of the economic and political weapons in the armoury of international creditors for use against poor debtor nations. There is little point in railing against the corporations themselves for this behaviour. Dogs bark, ducks quack and corporate capital rapaciously exploits the world's resources. Corporate Social Responsibility, about which we hear so much that it is often simply referred to as 'CSR', is about as likely as Vegetarianism for Tigers (VfT). But the use of our own money, taxpayers' money, in the furtherance of these interests by the European Union, its largely unelected and unaccountable institutions and the governments of its member states is a huge disgrace, an injustice which is within our power to correct. There are huge vested interests at stake, however, and sound arguments and rational persuasion will not bring about the world we want to see, one where children do not die for lack of a glass of clean water or a toilet. The only way to do so is to follow US labour leader Bill Haywood's advice. Issued about a century ago, it is still the only effective formula for change:

Don't get mad, get organised.

BIBLIOGRAPHY

Books

Alston, Margaret and Kent, Jenny (2004) *Social Impacts of Drought: A Report to NSW Agriculture* (Wagga Wagga: Centre for Rural Social Research, Charles Sturt University)

Andreu, Joaquin; Rossie, Giuseppe; Vagliasindi, Federico and Vela, Alicia (eds) (2006) *Drought Management and Planning for Water Resources,* (Boca Raton, Florida, USA: Taylor & Francis Group)

Ariès, Paul (2007) *Le Mésusage: Essai sur l'hypercapitalisme* (Lyon: Parangon/Vs)

Beauvais, Michael (2007) *Arroser Sans Gaspiller* (Paris: Rustica)

Belanyà, Belén; Brennan, Brid; Hoedeman, Olivier; Kishimoto, Satoko and Terhorst, Philipp (eds) (2005) *Reclaiming Public Water: Achievements, Struggles and Visions From Around the World* (Amsterdam: Corporate Europe Observatory/TransNational Institute)

Botterill, Linda Courtenay and Fisher, Melanie (eds) (2003) *Beyond Drought: People, Policy and Perspectives* (Collingwood, Victoria, Australia: CSIRO Publishing)

Bouguerra, Mohamed Larbi (2006) *Water Under Threat* (London and New York: Zed Books)

Brimblecombe, Sharon; Gallanaugh, Diana and Thompson, Catherine (eds) (1998) *The Hutchinson Encyclopedia of Science* (Oxford: Helicon Publishing Group)

Elhance, Arun P. (1999) *Hydro-Politics in the Third World: Conflict and Cooperation in International River Basins* (Washington, DC: United States Institute of Peace Press)

Friedman, Milton (1962) *Capitalism and Freedom* (Chicago: University of Chicago Press)

Hayek, F.A. (1960) *The Constitution of Liberty* (Chicago: University of Chicago Press)

— (1979) *Law, Legislation and Liberty Vol. 3 The Political Order of a Free People* (Chicago: University of Chicago Press)

Holland, Ann-Christin Sjölander (2005) *The Water Business: Corporations versus People* (London & New York: Zed Books)

Hunt, Constance Elizabeth (2004) *Thirsty Planet: Strategies for Sustainable Water Management* (London and New York: Zed Books)

Khan, Shahkrukh Rafi (ed.) (2002) *Trade and Environment: Difficult Policy Choices at the Interface* (London/New York/ Islamabad: Zed Books/Sustainable Development Policy Institute)

Krech, Shepard; McNeill, J.R. and Merchant, Carolyn (eds) (2003) *Encyclopaedia of World Environmental History* (New York: Routledge)

Lynas, Mark (2007) *Six Degrees: Our Future on a Hotter Planet* (London: Fourth Estate)

Marx, Karl and Engels, Friedrich (1875) *Critique of the Gotha Programme*, http://www.marxists.org/archive/marx/works/1875/gotha/index.htm

McCully, Patrick (2001) *Silenced Rivers: The Ecology and Politics of Large Dams, Enlarged and Updated Edition* (London and New York: Zed Books)

McKernan, Michael (2005) *Drought: The Red Marauder* (Crows Nest, New South Wales, Australia: Allen & Unwin)

Pearce, Fred (2006) *When the Rivers Run Dry: What Happens When Our Water Runs Out?* (London: Transworld Books)

van Apeldoorn, Bastiaan (2002) *Transnational Capitalism and the Struggle over European Integration* (London: Routledge)

Salman, S.A. and Bradlow, Daniel L. (2006) *Regulatory Frameworks for Water Resources Management: A Comparative Study* (Washington, DC: World Bank Publications)

Ward, Diana Raines (2002) *WaterWars: Drought, Flood, Folly, and the Politics of Thirst* (New York: Shepherd Books)

Official, Academic, Trade Union and NGO Reports

(All online references last accessed during December 2008)

Ackerman, Frank and Stanton, Elizabeth (2006) *Climate Change – the Costs of Inaction: Report to Friends of the Earth England, Wales and Northern Ireland* (Medford/Somerville, MA: Global Development and Environment Institute, Tufts University)

Bayliss, Kate (2001) *Water privatisation in Africa: Lessons from three case studies* (Public Services International Research Unit, University of Greenwich)

Belgium, Government of (2006) *Study on human rights and the access to water: Contribution of the Government of Belgium* (United Nations High Commissioner for Human Rights), http://www2.ohchr.org/english/issues/water/contributions/Belgium.pdf

Bernardini, Francesca (2007) *UNECE Convention on the Protection and Use of Transboundary Watercourses and International Lakes,*

'Cooperation through Joint Bodies', http://www.unece.org/env/water/partnership/part63.htm#632

Blueprint for Water, *Blueprint for Water: 10 steps to sustainable water by 2015*, http://www.blueprintforwater.org.uk/blueprint_0.html

Bouchier, Ian (1998) *Cryptosporidium in Water Supplies* (United Kingdom Department of the Environment, Transport and the Regions)

Brown, Oli (2007) *Climate change and forced migration: Observations, projections and implications* (United Nations Human Development Report Office Occasional Paper)

Bullock, P.; Jones, R.J.A. and Montanarella, L. (eds) (1999) *Soil Resources of Europe: European Soil Bureau Report No.6* (Luxembourg: Office for Official Publications of the European Union)

Burston, P. (2006) *Dry Rot: Is England's Countryside Dying of Thirst? The impacts of droughts and water shortages on England's wildlife* (Sandy, Beds: Royal Society for the Protection of Birds)

Carafa, Luigi (2008) *The Outlook for Compliance with EU Environmental Law: 'Pushing' for a Decentralised System of Complaints?* (Bruges: College of Europe, European Political and Administrative Studies)

Commission of the European Communities (CEC 2003) *Communication from the Commission to the Council, the European Parliament, the European Economic and Social Committee and the Committee of the Regions: Internal Market Strategy: Priorities 2003–2006* (Brussels: Commission of the European Communities)

— (CEC 2003a) *Communication from the Commission to the European Parliament and the Council: Towards sustainable water management in the European Union – First stage in the implementation of the Water Framework Directive 2000/60/EC* (Brussels: Commission of the European Communities)

— (CEC 2003b) *Communication from the Commission to the Council and the European Parliament – The Reform of State-Owned Enterprises in Developing Countries with focus on public utilities: the Need to Assess All the Options* (Brussels: Commission of the European Communities)

— (CEC 2003c) *Green Paper on Services of General Interest* (Brussels: Commission of the European Communities)

— (CEC 2004) *Commission Staff Working Paper: Fifth Annual Survey on the implementation and enforcement of Community environmental law 2003* (Luxembourg: Office for Official Publications of the European Communities)

— (CEC 2007) *A new EU Floods Directive*, http://ec.europa.eu/environment/water/flood_risk/index.htm

— (CEC 2007a) *Communication from the Commission to the European Parliament and the Council: Towards sustainable water management in the European Union – First stage in the implementation of the Water Framework Directive 2000/60/EC* (Brussels, 22/3/07 COM (2007) 128 final)

— (CEC 2007b) *Green Paper from the Commission to the Council, the European Parliament, the European Economic and Social Committee and the Committee of the Regions Adapting to climate change in Europe – options for EU action* (Brussels, 29/6/07, COM(2007) 354 final)

— (CEC 2007c) *Introduction to the new EU Water Framework Directive*, http://ec.europa.eu/environment/water/water-framework/info/intro_en.htm

— (CEC 2008) *Communication from the Commission to the Council and the European Parliament 2007 Environment Policy Review* (Brussels: Commission of the European Communities)

Commission of the European Communities Directorate General for Regional Policy (CEC-DG-REGI 2003) *Guide for Successful Public–Private Partnerships* (Brussels: European Commission)

Corporate Europe Observatory (CEO 2003) *Evian: Corporate Welfare or Water for All?* (CEO Info Brief 6)

— (CEO 2003a) *WTO and Water: The EU's Crusade for Corporate Expansion* (CEO Info Brief 3)

— (CEO 2007) 'Murky Water – PPIAF, PSEEF and other examples of EU aid promoting water privatisation'

Corporate Europe Observatory/FIVAS/Friends of the Earth International/World Development Movement (CEO 2007) *Media Briefing – Down the Drain: How aid for water privatisation could be better spent* (CEO/FIVAS/FoEI/WDM)

Cour des Comptes, France (CdCF 1997) *La Gestion des Services publics locaux d'Eau et d'Assainissement: Rapport au Président de la République suivi des Réponses des Administrations, Collectivités, Organismes et Enterprises* (Paris: Directions des Journaux Officiels)

De la Motte, Robin (2004) *D10h: WaterTime National Context Report – Netherlands* (Public Services International Research Unit, University of Greenwich)

Deltacommissie (2008) *Samen werken met water: Een land dat leeft, bouwt aan zijn toekomst: Bevindingen van de Deltacommissie 2008* (Rotterdam: Deltacommissie) Translations by the authors.

Deltawerken (2004a) 'The Flood of 1953' (Stichting Deltawerken Online) http://www.deltawerken.com/The-flood-of-1953/89.html

— (2004b) 'The Deltaworks' (Stichting Deltawerken Online) http://www.deltawerken.com/Deltaworks/23.html

Department of the Environment, Heritage and Local Government, Republic of Ireland (DEHLG-Ireland undated) *River Basin Management Projects*, http://www.wfdireland.ie/RBMP.htm

European Commission EuropeAid Co-operation Office Africa, Caribbean, Pacific Economic and trade cooperation (EuropeAid 2007) *Financing Proposal 9th EDF* (Brussels: Commission of the European Communities)

European Environmental Agency (EEA 2003) *Environmental issue report No 35: Mapping the impacts of recent natural disasters and technological accidents in Europe* (Copenhagen: European Environmental Agency)

— (EEA 2004) *Impacts of Europe's Changing Climate: An indicator-based assessment* (Luxembourg: Office for Official Publications of the European Communities)

European Environmental Bureau (EEB 2007) *Contribution to Informal Council discussion on Water Scarcity and Droughts, Lisbon, 30 August 2007*, http://www.eeb.org/activities/water/Water-Scarcity-Droughts-Lisbon-300807.pdf

— (EEB 2007a) *Media Briefing: 25 April 2007, NGOs respond to Parliament's Flood Risk vote*

— (EEB 2007b) 'MEPs stick to costly concrete to manage floods, say environmental NGOs' (Brussels/Strasbourg, 25 April 2007)

European Environmental Bureau and WWF (EEB/WWF 2006) *Making economics work for the environment: Survey of the economic elements of the Article 5 reports of the EU Water Framework Directive* (Brussels: European Environmental Bureau and WWF)

— (EEB/WWF 2008) *Letting the public have their say on water management: A snapshot analysis of Member States' consultations on water management issues and measures within the Water Framework Directive* (Brussels: European Environmental Bureau and WWF)

European Union Water Initiative (EUWI 2008) *Annual Report 2008*

Falkenmark, Malin; Fox, Patrick; Persson, Gunn and Rockström, Johan (2001) *Water Harvesting for Upgrading of Rainfed Agriculture: Problem Analysis and Research Needs* (Stockholm: Stockholm International Water Institute)

Goldemberg, José (ed.) (2000) *World Energy Assessment: Energy and the Challenge of Sustainability* (New York: United Nations Development Programme)

Green, Colin (2000) *The lessons from the privatisation of the wastewater and water industry in England and Wales* (Flood Hazard Research Centre, Middlesex University, England), www.meif.org/uk/document/download/green_berlinchap.doc

Gruppo Soges (2008) *Mid-term Evaluation of the Water Facility* (Gruppo Soges S.p.A Consortium)

Hadley Centre of the Met Office of the (UK) Department for Environment, Food and Rural Affairs (2005) *Climate change and the greenhouse effect: A briefing from the Hadley Centre* (London: Met Office)

Hall, David (1999) *Water and Privatisation in Latin America, 1999* (Public Services International Research Unit, University of Greenwich)

— (2001) *Water Privatisation: Global Problems, Global Resistance* (Public Services International Research Unit, University of Greenwich)

— (2001a) *Water privatisation and quality of service – PSIRU evidence to the Walkerton enquiry, Toronto, July 2001* (Public Services International Research Unit, University of Greenwich)

— (2003a) *Water and DG Competition* (Public Services International Research Unit, University of Greenwich)

— (2003b) *EC Internal market strategy – implications for water and other public services* (Public Services International Research Unit, University of Greenwich)

— (2003c) *A critique of the EC green paper on Services of General Interest* (Public Services International Research Unit, University of Greenwich)

— (2004) *Privatising other people's water – the contradictory policies of Netherlands, Norway and Sweden* (Public Services International Research Unit, University of Greenwich)

— (2006) *Water and electricity in Nigeria* (Public Services International Research Unit, University of Greenwich)

— (2008) *EU Neighbourhood policy: implications for public services and trade unions* (Public Services International Research Unit, University of Greenwich)

Hall, David; Bayliss, Kate and Lobina, Emanuele (2001) *Still fixated with privatisation: A Critical Review of the World Bank's Water Resources Sector Strategy*, paper prepared for the International Conference on Freshwater, Bonn, Germany, 3–7 December 2001 (Public Services International Research Unit, University of Greenwich)

— (2002) *Water privatisation in Africa*, paper presented at Municipal Services Project Conference, Witswatersrand University, Johannesburg May 2002 (Public Services International Research Unit, University of Greenwich)

Hall, David; Lobina, Emanuele and de la Motte, Robin (2003) *Water privatisation and restructuring in Central and Eastern Europe and NIS countries, 2002* (Public Services International Research Unit, University of Greenwich)

— (2004a) *Making water privatisation illegal: new laws in Netherlands and Uruguay* (Public Services International Research Unit, University of Greenwich)

Hall, David; Corral, Violeta; Lobina, Emanuele and de la Motte, Robin (2004) *Water privatisation and restructuring in Asia–Pacific* (Public Services International Research Unit, University of Greenwich)

Hall, David and Lobina, Emanuele (2005) *The relative efficiency of public and private sector water* (Public Services International Research Unit, University of Greenwich)

— (2006) *Pipe dreams: The failure of the private sector to invest in water services in developing countries* (World Development Movement)

— (2006a) *Water as a public service* (Public Services International Research Unit, University of Greenwich)

— (2007) *Water companies in Europe 2007* (Public Services International Research Unit, University of Greenwich)

Hukka, Jarmo J. and Seppälä, Osmo T. (2005) *D22: WaterTime case study – Hämeenlinna, Finland* (Water Time Case Study reports, Public Services International Research Unit, University of Greenwich)

House of Commons Select Committee on the Environment Seventh Report, *1999–2000: Water Prices and the Environment HC 597* (HOCSC 2000), 14 November 2000

Intergovernmental Panel on Climate Change (IPCC 1995) *Second Assessment Report* (Cambridge, England: Cambridge University Press)

— (IPCC 2008) *Technical Paper on Climate Change and Water* (Geneva: IPCC)

Kuneman, Gijs (2002) *Greener Fields: EEB Position Paper on further reform of the Common Agriculture Policy* (Brussels: European Environmental Bureau)

Lanz, Klaus and Scheuer, Stefan (2001) *EEB Handbook on EU Water Policy under the Water Framework Directive* (Brussels: European Environmental Bureau)

Lariola, M. (2000) *Twinning cooperation between Riga Water Company and Stockholm Water Company* (Stockholm: Sida Evaluations, 2000/7)

Lariola, M. and Danielsson, B. (1998) *Twinning cooperation between Kaunas Water Company, Lithuania and Stockholm Water Company* (Sida Evaluations, 1998/19), (Stockholm: Sida)

Lobina, Emanuele (2000) *Grenoble – water re-municipalised* (London: PSIRU, University of Greenwich)

— (2002) *Water privatisation and restructuring in Central and Eastern Europe, 2001* (Public Services International Research Unit, University of Greenwich)

Lobina, Emanuele and Hall, David (1999) *Public Sector Alternatives to Water Supply and Sewerage Privatisation: Case Studies* (London: PSIRU, University of Greenwich)

— (2001) *Water privatisation – a briefing* (London: PSIRU, University of Greenwich)

— (2006) *Public–Public Partnerships as a catalyst for capacity building and institutional development: Lessons from Stockholm Vatten's experience in the Baltic region* (Public Services International Research Unit, University of Greenwich)

Lundqvist, J.; de Fraiture, C. and Molden, D. (2008) *Saving Water: From Field to Fork, Curbing Losses and Wastage in the Food Chain* (Stockholm: Stockholm International Water Institute)

Miller, Claire (ed.) (2006) *Conflicting EU funds: Pitting conservation against unsustainable development* (Vienna: WWF Global Species Programme)

Mohajeri, Shahrooz; Knothe, Bettina; Lamothe, David-Nicolas and Faby, Jean-Antonie (eds) (2003) *European Water Management between Regulation and Competition* (Luxembourg: Office for Official Publications of the European Communities)

Moren-Abat, Marta et al. (2007) *International Workshop: Climate Change Impacts on the Water Cycle, Resources and Quality Research-Policy Interface, 25–26 September 2006*, Scientific and policy report (Brussels, Commission of the European Communities Directorate-General for Research/Luxemourg: Office for Official Publications of the European Communities)

Montana, State of (2004) *The State of Montana Multi-Hazard Mitigation Plan and Statewide Hazard Assessment* (Fort Harrison, MT, State of Montana)

Myers, Norman (2005) 'Environmental Refugees: An emergent security issue' (Organisation for Security and Cooperation in Europe, Paper to 13th Economic Forum, May 2005, Prague)

Organisation for Economic Cooperation and Development (OECD 2001) *OECD Environmental Strategy for the First Decade of the 21st Century Adopted by OECD Environment Ministers 16 May 2001*, http://www.oecd.org/dataoecd/33/40/1863539.pdf

— (OECD 2004) *The OECD Environmental Strategy: Progress in Managing Water Resources* (Paris: OECD)

— (OECD 2005) *Development Assistance Committee (DAC), Issues Brief: Maintreaming Conflict Prevention: Water and Violent Conflict* (Paris: OECD)

Public Citizen (2003) *Water Privatization Fiascos: Broken Promises and Social Turmoil* (Washington, DC: Public Citizen Water for All program)

Public Services International Research Unit (PSIRU 2000) *It Cannot be Business as Usual: Problems with the Private Models for Water* (Public Services International Research Unit, University of Greenwich) www. psiru.org/reports/2000-03-W-Hmodel.doc

— (PSIRU 2000a) *The truth about privatisation* (Public Services International Research Unit, University of Greenwich)

Steenblik, Ronald (2007) *Biofuels – At What Cost? Government support for ethanol and biodiesel in selected OECD countries* (Geneva: The Global Subsidies Initiative (GSI) of the International Institute for Sustainable Development (IISD))

Swenson, Dave; Eathington, Liesl and O'Brien, Meghan (2008) *Economic Impacts of the 2008 Floods in Iowa* (Ames, Iowa: Iowa State University Regional Capacity Analysis Program, 2008) http://www.econ.iastate. edu/research/webpapers/paper_12954.pdf

Tearfund/Water Aid (2006) *An empty glass: The EU Water Initiative's contribution to the water and sanitation Millennium targets* (Tearfund/ Water Aid)

Thomas, Stephen and Hall, David (2006) *GATS and the electricity and water sector* (Public Services International Research Unit, University of Greenwich)

UNICEF (2002) *Meeting the MDG Drinking Water and Sanitation Target: A Mid-Term Assessment of Progress, 2002,* http://www.unicef. org/wes/mdgreport/index.php

United Nations (UN 1978) *United Nations Conference on Desertification, 29 August – 9 September 1977. Round-up, Plan of Action and Resolutions* (New York, United Nations) available at http://www. ciesin.org/docs/002-478/002-478.html

— (UN 2003) *The World's Water Crisis* (Geneva: United Nations)

— (UN 2006) *World Water Development Report: Water for People, Water for Life* (Oxford: UNESCO & Berghan Books)

United Nations Department of Economic and Social Affairs (UN-DESA 2008) *World Urbanization Prospects: The 2007 Revision, Executive Summary* (New York: United Nations)

United Nations Development Programme (UNDP 2006) *Human Development Report 2006: Beyond scarcity: Power, poverty and the global water crisis* (New York: UNDP)

United Nations Economic Commission for Europe (UNECE 2004) *The 1992 UNECE Convention on the Protection and Use of Transboundary Watercourses and International Lakes* (New York and Geneva: United Nations)

United Nations Environmental Programme (UNEP 1992) *Agenda 21: Protection of the quality and supply of freshwater resources: Application of integrated approaches to the development, management and use*

of water resources, http://www.unep.org/Documents.Multilingual/
Default.Print.asp?DocumentID=52&ArticleID=66

— (UNEP 1999) *Global Environment Outlook 2000* (Nairobi: UNEP)

Water Framework Directive Common Implementation Strategy (WFDCIS
2001)*Water Framework Directive Common Implementation Strategic
Document as agreed by the Water Directors under Swedish Presidency*,
http://ec.europa.eu/environment/water/water-framework/objectives/
pdf/strategy.pdf

Water Information System for Europe (WISE undated) *Water Note 5
Economics in Water Policy: The value of Europe's waters* (Brussels:
European Commission DG Environment)

Waterland – Water in the Netherlands (WWIN) *Flood Control and
Protection*, http://www.waterland.net/index.cfm/site/Water%
20in%20the%20Netherlands/pageid/E3B3B416-FB4E-0AB8-
2FB6E2B271F1BD6E/index.cfm

Winpenny, James (2003) *Financing Water for All: Report of the World
Panel on Financing Water Infrastructure chaired by Michael Camdessus*
(Kyoto: 2003 World Water Forum)

Wolf, Aaron T. (UNEP 2002) *Atlas of International Freshwater
Agreements* (Nairobi: United Nations Environmental Programme/
Food and Agriculture Organisation)

World Commission on Dams (WCD 2000) *Dams and Development:
A New Framework for Decision-Making – The Report of the
World Commission on Dams* (London and Sterling, Va: Earthscan
Publications)

World Economic Forum (WEF 2008) *Managing our Future Water Needs
for Agriculture, Industry, Human Health and the Environment:
Discussion Document for the World Economic Forum Annual Meeting
2008* (Geneva: WEF)

WHO/UNICEF Joint Monitoring Programme for Water Supply and
Sanitation (WHO 2000) *Global Water Supply and Sanitation
Assessment 2000 Report*, http://www.who.int/docstore/water_
sanitation_health/Globassessment/GlobalTOC.htm

WRc/Ecologic (2002) *Study on the application of competition rules to the
water sector in the European Community* (Brussels: Commission of the
European Communities – Directorate General for Competition)

WWF Policy Briefing (WWF 2004a) *Living With Floods: Achieving
ecologically sustainable flood management in Europe* (Brussels: WWF
European Policy Office)

WWF (WWF 2004b) *Rivers at Risk: Dams and the future of freshwater
ecosystems* (Godalming, England: Dam Right! WWF Dams Initiative,
2004)

Yoffe, Shira; Wolf, Aaron T. and Giordano, Mark (2003) *Conflict and Cooperation over International Freshwater Resources* (Internet Publication, University of Oregon), http://www.transboundarywaters. orst.edu/research/basins_at_risk/bar/

Papers, Articles, Pamphlets and Presentations

(All online references last accessed during December 2008)

Bailey, Robert (2008) *Bio-fuelling Poverty? Feeding the World under the Climate Threat* (Powerpoint presentation by Oxfam senior policy adviser, Conference on 'Feeding the world under the climate threat', European Parliament, Brussels, 12 November 2008)

Bernardini, Francesca (2007) *Convention on the Protection and Use of Transboundary Watercourses and International Lakes* (United Nations Economic Commission for Europe), http://www.inbo-news. org/ag2007/comms/7_16_45/bernardini_UNECE_Debrecen.pdf

Chong, Eshien; Huet, Freddy; Saussier, Stéphane and Steiner, Faye (2006) 'Public–Private Partnerships and Prices: Evidence from Water Distribution in France', *Review of Industrial Organization*, 29, 1–2, September 2006

Cooper, J.R. and Gilliam, J.W. (1987) 'Phosphorus Redistribution from Cultivated Fields into Riparian Areas,' *Soil Science Society of America Journal*, 51, pp.1600–1604

Cooper, J.R.; Gilliam, J.W.; Daniels, R.B. and Robarg, W.P. (1987) 'Riparian areas as filters for agricultural sediment', *Proceedings of the Soil Society of America*, 51, pp.416–20

D'Silva, Joyce (2008) *The impact of increased meat consumption on developing countries* (Compassion in World Farming, Powerpoint presentation, Conference on 'Feeding the world under the climate threat', European Parliament, Brussels, 13 November 2008)

Evans, Robert; Gilliam, J.W. and Lilly, J.P. (1996) *Wetlands and water quality* (Winston–Salem, NC: North Carolina Cooperative Extension Service), http://www.bae.ncsu.edu/programs/extension/evans/ag473-7.html

Fuentelsaz, Felipe (2008) *Reducing water wastage: irrigation in the Donana Delta* (WWF Spain; Powerpoint presentation at EEB Conference, 'Crunch time for Europe's water', Brussels, 18 November 2008)

Gill, Stephen (1995) 'Globalisation, Market Civilisation, and Disciplinary Neoliberalism', *Millennium: Journal of International Studies*, 24(3), pp.399–423

— (1998) 'New Constitutionalism, Democratisation and Global Political Economy', *Pacifica Review*, 10(1), pp.23–38

Gleick, Peter H. (1996) 'Basic Water Requirements for Human Activities: Meeting Human Needs', *Water International*, 21, pp.83–92

— (1997) 'Water Planning and Management Under Climate Change', undated on the website of the American Water Works Association which notes, however, that 'Portions of this article appeared in a piece prepared by Peter H. Gleick for the Journal of the American Water Works Association ("Climate Change and Water Resources") in November 1997, and in Chapter 5 of the new book *The World's Water 1998–1999* (Island Press, Washington, D.C.)'

— (1999) 'The Human Right to Water', *Water Policy*, 1(5), pp.487–503, available online at http://www.pacinst.org/reports/basic_water_needs/human_right_to_water.pdf

— (2000) *Water Conflict Chronology* (Pacific Institute for Studies in Development, Environment, and Security), http://www.worldwater.org/conflictchronologychart.PDF

Godoy, Julio (2003) 'Water and Power: The French Connection' (Washington DC: Center for Public Integrity), http://globalpolicy.igc.org/socecon/tncs/2003/0204frenchwater.htm

Green, Colin (2005) *The battle against water privatisation* (Dublin: Socialist Democracy)

Gustafsson, Jan-Erik (2004) 'Political fashion has driven the privatisation of public utilities in Sweden', *European Business Forum*, 18, Summer 2004, http://www.ebfonline.com/Article.aspx?ArticleID=243

Hall, David; Lobina, Emanuele and de la Motte, Robin (2005) 'Public resistance to privatisation in water and energy', *Development in Practice*, Vol.15, 3 & 4, June 2005

Herrero, Amaranta (2008) *Socio-environmental impacts of agrofuels: Lessons from sugarcane ethanol production* (Corporate Europe Observatory, Powerpoint Presentation, Conference on 'Feeding the world under the climate threat', European Parliament, Brussels, 12 November 2008)

Hoedeman, Olivier (2007) 'European Double Standards', *TransNational Institute*, March 2007, http://www.tni.org/detail_page.phtml?&act_id=16533&menu=05k

International Study Circles 2nd Pilot, March 1998 – June 1998 (ISC 1995) *Case-study: Lithuania: Union defeats TNC*, http://www.ifwea.org/isc/pilot02/ses3/read1.html

Jackson, M.B. and Colmer, T.D. (2005) 'Response and Adaptation by Plants to Flooding Stress', *Annals of Botany*, 96(4), pp.501–5

Maurits la Rivière, J.W. (1989) 'Threats to the world's water', *Scientific American*, 261(3), September 1989, pp.80–94

Orwin, Alexander (1999) *The Privatization of Water and Wastewater Utilities: An International Survey* (Brunswick, Ontario: Environment Probe) http://www.environmentprobe.org/EnviroProbe/pubs/ev542.html

Pietilä, Pekka (2005) 'Role of Municipalities in Water Services in Namibia and Lithuania', *Public Works Management & Policy*, 10(1), pp.53–68

Sadoff, Claudia W. and Grey, David (2002) 'Beyond the River: the benefits of cooperation on international rivers', *Water Policy*, 4, pp.389–403

Sparks, Richard E. (2006) 'Rethinking, Then Rebuilding New Orleans', *Issues in Science and Technology*, Winter 2006

Van Gelder, Pieter H.A.J.M. (1999) *Risks and safety of flood protection structures in the Netherlands* (Delft: Technical University of Delft) http://www.aia.org/SiteObjects/files/Risks%20and%20safety%20of%20flood%20protection.pdf

Vinnari, Eija M. and Hukka, Jarmo J. (2007) 'Great expectations, tiny benefits – Decision-making in the privatization of Tallinn water', *Utilities Policy*, 15(2), pp.78–85

Weingast, Barry (1995) 'The Economic Role of Political Institutions: Market-Preserving Federalism and Economic Development', *Journal of Law, Economics, & Organization*, 11(1), pp.1–31

NOTES

Introduction: A Dangerous Synergy

1. For a fuller discussion of the range of human uses of water, see Hunt 2004, especially pp.44–8.
2. UN 2003, p.12.
3. Bailey 2008.
4. UN 2006, p.4.
5. Bouguerra 2006, p.51.
6. Bouguerra 2006, p.11.
7. Vandana Shiva, 'Why we face both food and water crises', interview with *Alternet,* 15 May 2008, http://www.alternet.org/environment/85433/
8. Shiney Varghese, 'The Triple Threat: Our Food, Water and Climate Challenges', 14 May, 2008, http://www.alternet.org/environment/85414/
9. This information, and all of the strictly scientific information used in this book comes from a variety of sources which the authors, as non-scientists with a keen lay interest in the sciences, have been obliged to rely on. In this case the principal source is Michael Beauvis, *Arroser sans gaspiller* (Paris: Editions Rustica, 2007), essentially a practical book for gardeners, and Ian Woodward, 'Plants in the Greenhouse World', *New Scientist,* 6 May 1989, http://www.newscientist.com/article/mg12216636.800-plants-in-the-greenhouse-world.html
10. National Drought Mitigation Center (USA), 'Impacts of Drought', http://www.drought.unl.edu/risk/impacts.htm
11. UN 2003, p.12.
12. A comprehensive list of extreme weather events involving unusual levels of precipitation is on the National Climatic Data Center of the US Department of Commerce at http://lwf.ncdc.noaa.gov/oa/climate/severeweather/rainfall.html#2008. Although somewhat out of date, the discussion 'Are Recent Extreme Weather Events, Like the Large Number of Atlantic Hurricanes in 1995, Due to Global Warming?' published by the United Nation's Environmental Programme in 1997 and available at http://www.gcrio.org/ipcc/qa/08.html remains useful.
13. IPCC 2008, p.52.

14. See Lynas 2007, Chapter 2, for a detailed discussion of what 2° warming would be likely to mean.
15. Vicky Pope, 'What can climate scientists tell us about the future?', 2 May 2008, (Dr Pope is described as Head of Climate Change for Government, Met Office Hadley Centre) http://www.metoffice.gov. uk/research/hadleycentre/future/
16. Myers 2005, p.1.
17. Brown 2007, p.7.
18. Myers 2005, p.3.
19. Janos Bogardi, quoted in communiqué from United Nations University (2005) 'As Ranks of "Environmental Refugees" Swell Worldwide, Calls Grow for Better Definition, Recognition, Support'. Bogardi is described as Director of the UN University Institute for Environment and Human Security. The communiqué concerned the UN Day for Disaster Reduction, 12 October 2005. Cited in Brown 2007, p.25.
20. Bouguerra 2006, p.126.
21. UN 2003, p.11; adequate access is defined as 'the availability of at least 20 l. of water per person per day from an improved source within 1 km., a very minimalist definition indeed'. IPCC 2008, p.87.
22. Bouguerra 2006, p.128.
23. See the European Commission's water policy portal, http://ec.europa. eu/environment/water/index_en.htm
24. Cited in UNEP 1999, http://www-cger.nies.go.jp/geo2000/english/ text/0078.htm 'Chapter Two: The State of the Environment – Europe and Central Asia'.
25. Ariès 2007, translation by S. McGiffen.
26. *Ibid.*

1 Drought and Deprivation

1. McKernan 2005, pp.3–4.
2. Janette A. Lindesay, 'Climate and Drought in Australia' in Botterill and Fisher 2003, p.38.
3. IPCC 2008, pp.46–7.
4. Javier Ferrer Polo and Javier Paredes, 'Water management in Mediterranean regions prone to drought: The Jucar Basin experience', p.2, in Andreu 2006. Ferrer and Paredes also repeat and endorse the IPCC's definition, though for some reason they leave out 'Environmental Drought' and include 'Socio-economic drought' as their fourth factor.
5. Botterill and Fisher 2003, p.3.

6. J. Andreu and A. Solana, 'Methodology for the analysis of drought mitigation measures in water resource systems', in Andreu 2006, pp.134 and 135.

7. UNDP 2006, pp.211–13.

8. Gleick 1996.

9. Hunt 2004, p.50.

10. The definitions offered by Malin Falkenmark, Senior Scientist at the Stockholm International Water Institute (SIWI), have become generally accepted. Water scarcity is suffered, according to these definitions, by countries with a potential annual water supply of more than 1000 cubic metres per person and less than 1700; water stress is suffered by those having more than 500 and less than 1000 cubic metres. Any country with less than 500 cubic metres per person per year is 'in crisis'. See Hunt 2004, p.48.

11. Hunt 2004, p.49.

12. UN 2003, p.10.

13. Karl Russel, 'Populations Are Expanding Fastest in Regions Where it Is Most Difficult to Grow Food', *New York Times*, 21 July 2008.

14. UNDP 2006, pp.23–4.

15. A distinction is made generally between 'consumptive use', in which the water used is not returned to the supply until it returns to it naturally through the hydrological cycle (e.g. watering crops) and uses of water where the water may be automatically or easily recycled (e.g. cooling).

16. Tuber-eaters such as the authors are the major exception, the Netherlands and the north of England being amongst the world's leading potato consumers.

17. WEF 2008, p.4.

18. D'Silva 2008.

19. Lundqvist et al 2008, p.10.

20. Lundqvist et al 2008, p.11.

21. Lundqvist et al 2008, p.25.

22. Bouguerra et al 2006, p.96.

23. Daniel Loucks, 'Decision support systems for drought management', in Andreu 2006, pp.119–20.

24. South Australia Murray Irrigators (SAMI), 'Over-allocation now a global phenomenon: A special report about river management worldwide', SAMI News, July 2008, http://www.samurrayirrigators.org.au/sami_news.asp

25. See Andres Sahuquillo, 'Strategies for the conjunctive use of surface and groundwater', in Andreu et al 2006, pp.49–71. The 'Alternative' in Alternative Conjunctive Use refers to the contrast with the more

common, more conventional, but more expensive practice of artificial replenishment of underground aquifers from surface waters.

26. Andreu et al 2006, pp.23–4.
27. IPCC 2008, p.43. The exception mentioned is southwestern Australia, where 'increased groundwater withdrawals have not only been caused by increased water demand but also because of climate-related decrease in recharge from surface water supplies'.
28. Hunt 2004, p.57.
29. UN-DESA 2008, p.11.
30. UN-DESA 2008, p.3.
31. UN 2003, p.14.
32. UN 2003, p.10.
33. Hunt 2004, p.104.
34. Hunt 2004, p.114.
35. Tariq Banuri, 'Alternative Public Regimes for Achieving Environmental Improvement in the Global Cotton Commodity Chain: The Case of Pakistan', Chapter 3 of Khan 2002, pp.74–5. This essay, and Chapter 4 on 'Environmental impacts and mitigation costs', contain interesting proposals for reducing pollution from the Pakistani cotton and textile industries.
36. Hunt 2004, pp.104–5.
37. Shahrukh Rafi Khan, Mahmood A. Khwaja, Abdul Matin Khan, Haider Ghani and Sajid Khazm, 'Environmental Impacts and Mitigation Costs: The Case of Pakistan's Cloth and Leather Exports', Chapter 4 of Khan 2002, p.113.
38. See Bouguerra 2006, Chapter 10, 'Pollution in many different forms', pp.142–56.
39. IPCC 2008.
40. For a brief summary overview of the environmental impacts of drought, see 'Understanding Your Risk and Impacts: Impacts of Drought: Environmental Impacts', at http://drought.unl.edu/risk/environment.htm
41. 'An additional hazard resulting from drought conditions is insect infestation. In the Northern Great Plains, rangeland grasshopper outbreaks have caused significant damage to the agricultural economy. Grasshopper populations tend to increase with both livestock grazing rates and dry conditions, and they can double, triple, or quadruple with each successive year of drought.' State of Montana 2004, Section 3, 3.3.8, 'Weather – Drought and Effects of Drought', p.95.
42. This information has been gathered in various places, including in conversation with scientists and environmental activists, but two books have been used to check details: *The Hutchinson Encyclopedia*

of Science (Oxford: Helicon, 1998) and Michael Beauvais *Arroser Sans Gaspiller* (Paris: Rustica, 2007).

43. Burston 2006, pp. 4–7.

44. McKernan 2005, pp.10–11.

45. Alston and Kent 2004, p.1. The other group of researchers referred to here are those in Botterill 2003.

46. Alston and Kent 2004, p.57.

47. *Ibid.* Interviews were reported anonymously with individuals identified only by occupation or function.

48. Alston and Kent 2004, pp.94–5.

49. 'Drought Plagues Australia', *Echolist* (Australian on-line news source), 16 August 2007, http://www.echolist.com/world/2007/august/news272627.html

50. Alston and Kent 2004.

51. Quoted in Andrew Heavens 'In Ethiopia, schools empty as effects of drought wear on', 29 June 2006, http://www.unicef.org/infobycountry/ethiopia_34733.html

52. Augustine Agu, UNICEF Ethiopia Education Chief, quoted in 'In Ethiopia, schools empty as effects of drought wear on', *UNICEF Newsline*, 29 June 2006, http://www.unicef.org/infobycountry/ethiopia_34733.html

53. UNICEF, *Ethiopia – Background,* http://www.unicef.org/infobycountry/ethiopia_12162.html

54. OECD-DAC 2005, p.3.

55. Quoted in Bouguerra 2006, p.79.

56. Andreu et al 2006, p.121.

57. Daniel P. Loucks, in Andreu 2006, Chapter 5, 'Decision support systems for drought management', p.13–132.

58. 'Conservations and Rebates', Southern Nevada Water Authority's website, http://www.snwa.com/html/cons_index.html

59. South Australia Murray Irrigators (SAMI), 'Over-allocation now a global phenomenon: A special report about river management worldwide', SAMI News, July 2008, http://www.samurrayirrigators.org.au/sami_news.asp. The facts are taken from SAMI's report. The sceptical views in relation to the Governor's proposals are the authors' own.

60. Hunt 2004, p.101.

61. UN 1978; Hunt 2004, pp.101–3.

62. For a brief account of the related processes of salinisation and waterlogging, see WCD 2000, pp.66–8.

63. Hunt 2004, pp.101–3.

64. Bullock et al 1999.

2   Flood

1. Blues song quoted in the film *Fatal Flood* (United States Public Broadcasting Service, 1999), from the series *American Experience*.
2. Andreu et al 2006, p.2.
3. Sean Poulter, 'A million homes built on flood plains could be denied insurance', *Daily Mail*, 9 July 2007; Tania Branigan, 'Houses can be built on flood plains, minister insists', *The Guardian*, 24 July 2007.
4. WWIN, 'Flood Control and Protection'.
5. UN 2003, p.12; Bouguerra 2006, p.76.
6. IPCC 2008, p.19.
7. Hunt 2004, p.135.
8. OECD 2004, p.5.
9. Jackson and Colmer 2005.
10. Amanda Lee Myers, 'Three-Day Grand Canyon Flood Aims to Restore Ecosystem', *Associated Press* in *National Geographic*, 6 March 2008.
11. Sparks 2006.
12. WWF 2004a, p.6.
13. Hunt 2004, p.133.
14. IPCC 2008, p.89.
15. IPCC 2008 p.51.
16. Swenson et al 2008.
17. Deltawerken 2004a.
18. Sparks 2006.
19. 'Gustav slams La. coastline west of New Orleans', AP report, 1 September 2008, http://apnews.myway.com/article/20080901/D92U0LJO2.html
20. Sparks 2006.
21. Hunt 2004, p.136.
22. Hunt 2004, pp.138–9.
23. Hunt 2004, p.139.
24. Hunt 2004, p.141.
25. Hunt 2004, pp.141–2.
26. Hunt 2004, p.142.
27. Van Gelder 1999, p.55.
28. Deltawerken 2004b.
29. Van Gelder 1999, pp.57–8.
30. Deltacommissie 2008.
31. Deltacommissie 2008, p.9.
32. Deltacommissie 2008, p.10.

33. To declare an interest, the authors are respectively Member of the European Parliament (Liotard) for, and translator and consultant (McGiffen) to, the same party.

34. Quotes from Paulus Jansen come from personal communications and from Socialist Party of the Netherlands Communiqué at http://www.sp.nl/milieu/nieuwsberichten/5938/080903-verbreed_deltawet_tot_klimaatwet.html, 'Verbreed Deltawet tot Klimaatwet', 3 September 2008.

35. Quoted in 'Wetenschappers bekritiseren Deltacommissie om worstcasescenario', *NRC Handelsblad* 9/10/08, translation by S. McGiffen.

36. WCD 2000, pp.58–60.

37. Hunt 2004, p.142.

38. WCD 2000, p.60.

39. Ward 2002, p.11.

40. WCD 2000, p.60.

41. *Ibid.*

42. WCD 2000, pp.60–1.

43. Hunt 2004, pp.144–5.

44. Pearce 2006, p.157; for a thorough account see the table, 'Cost and time overruns for dam projects', pp.266–7; McCully 2001.

45. Pearce 2006, pp.157–9.

46. WCD 2000, p.61.

47. *Ibid.*

48. WCD 2000, p.62.

49. Hunt 2004, p.147.

50. *Ibid.*

51. Hunt 2004, pp.148–9.

52. Hunt 2004, pp.149–50.

53. Hunt 2004, pp.153–4; Pearce 2006, pp.321–31.

3 Conflict and Cooperation

1. Directed by Roman Polanski and starring Jack Nicholson and Faye Dunaway, *Chinatown* (1974) centres around a dispute over water supply between Los Angeles and nearby rural areas. The events depicted in the film, though it is set in the 1930s, bear a close resemblance to a real dispute which occurred some 30 years earlier.

2. UNDP 2006, p.205; see also Table 6.1, p.206.

3. OECD-DAC 2005, p.3.

4. For a brief account of this longstanding conflict, see Isabelle Humphries, 'Breaching Borders: The Role of Water in the Middle East Conflict', Sept/Oct 2006, pp.20–1.

5. Ashley Powdar, 'One of History's Great Atrocities: The Corporate Theft of the Public's Right to Water', Council on Hemispheric Affairs, 2008, http://www.coha.org/2008/04/one-of-history%e2%80%99s-great-atrocities-the-corporate-theft-of-the-public%e2%80%99s-natural-right-to-water/; Marcela Valente, 'Specter of Water War Looms Over Guaraní Aquifer', *Tierramerica*, 2004, http://www.tierramerica.net/english/2004/0320/iarticulo.shtml; Krech et al 2003, p.706.

6. UNDP 2006, pp.204–5.

7. Anil Naidoo and Adam Davidson-Harden, 'Water as a strategic international resource: The new water wars', http://www.cancun2003.org/download_en/Naidoo_Davidson-Harden_conflict_water_version_final_ingles.doc

8. Yoffe et al 2003, p.101.

9. Yoffe et al 2003, p.102.

10. UNDP 2006, p.221.

11. Elhance 1999, p.6.

12. Raphaele Bail, 'L'eau: Comment mieux la répartir', *Faim Développement Magazine*, No 207, November 2005, p.16; see Gleick 2000 for an exhaustive chronology of water-based armed conflicts since 1503.

13. Bouguerra 2006, pp.66–7.

14. Powerpoint presentation by Dr. Miklós Varga, János Rémai and Kálmán Papp, 'Transboundary Cooperation, Hungary', undated, *http://riob.org/divers/thonon/hung_angol.PDF*

15. Sadoff and Grey 2005, p.399.

16. Antonia Juhasz, 'The Corporate Invasion of Iraq', International Forum on Globalization website, http://www.ifg.org/analysis/globalization/iraqinvasion.html, taken from *Left Turn* magazine, Aug/Sept 2003.

17. OECD-DAC 2005, pp.1–2

18. OECD-DAC 2005, p.3.

19. OECD-DAC 2005, p.4.

20. OECD-DAC 2005, p.5.

21. Elhance 1999, p.109.

22. OECD-DAC 2005, p.6.

23. Sadoff and Grey 2002, p.399.

24. Sadoff and Grey 2002, p.400.

25. UNDP 2006, p.210.

26. Sadoff and Grey 2002, p.400.

27. Elhance 1999, p.224.

28. UNDP 2006, pp.217–18.

29. UNDP 2006, p.217.

30. Lonergan, Stephen C. and Brooks, David B. (1995) *Watershed: The Role of Fresh Water in the Israeli-Palestinian Conflict* (Ottawa: International Development Research Centre). Appendix 2, 'Declarations, documents, rules, and rhetorics: Fresh water as a global issue'.

31. *Ibid.*

32. *The Guiding Principles from the Dublin Statement (1993) on water and sustainable development,* http://www.wmo.ch/pages/prog/hwrp/documents/english/icwedece.html#principles

33. UNDP 2006, p.218.

34. Sadoff and Grey 2002, p.397.

35. Wolf 2002, p.5; Bernardini 2007; see also UNECE 2004 for summary of Convention; and http://www.unece.org/env/water/text/text.htm for full text.

36. Bernadini 2007.

37. *Ibid.*

38. For a fuller definition and discussion of the Precautionary principle, see 'Conference on the Precautionary Principle, January 26, 1998', Science and Environmental Health Network, http://www.sehn.org/wing.html

39. Bernardini 2007.

40. Bernardini 2007, UNECE Convention on the Protection and Use of Transboundary Watercourses and International Lakes , 'Cooperation through Joint Bodies', http://www.unece.org/env/water/partnership/part63.htm#632

41. UNDP 2006, p.222.

42. See Wolf 2002, p.8, for recommendations which 'may assist the international, regional and basin communities as they expand and refine their cooperative water structures'.

43. UNDP 2006, p.223.

44. Elhance 1999, p.225.

45. Elhance 1999, p.226.

46. Elhance 1999, pp.233–8.

4   It Never Rains But it Pours: Climate Change, Water Shortage and Flood

1. IPCC 2008, p.15; pp.34–5.

2. Lynas 2007, pp.19–20.

3. IPCC 2008, p.39; p.44.

4. EEA 2004, p.65.

5. Lynas 2007, pp.132–6.

6. Lynas 2007, p.59.

7. IPCC 2008, p.49.

8. Hunt 2004, p.198.

9. IPCC 1995, p.133, cited in Hunt 2004.

10. 'Evapotranspiration is a function of the climactic demand for water and the supply of water. It is driven by energy availability, particularly net radiation. An increase in net radiation increases the demand for evaporation. The effect is complicated by changes in atmospheric humidity, which alter the capacity of the air to accept more water, and by changes in the rate of air movement across evaporating surfaces. The humidity of air masses is also linked to evaporation over land and water bodies, including the ocean. Higher temperatures will increase the capacity of the atmosphere to hold water, thus enhancing the effects of increased net radiation where air humidity currently imposes a constraint on the rate of evaporation, as in humid areas. A temperature rise of about two degrees... could cause an increase in potential evaporation of up to 40 percent in a humid temperate region, but less in a drier environment, where changes in humidity are not as important in determining evapotranspiration rates.' Hunt 2004, pp.197–8.

11. Hunt 2004, p.198.

12. IPCC 2008, p.54.

13. Hunt 2004, p.199.

14. Wetlands International, 'Threatened wetlands', http://www.wetlands. org/Aboutwetlandareas/Threatenedwetlandsites/tabid/1125/ language/en-US/Default.aspx

15. IPCC 2008, p.49.

16. Wetlands International, 'Threatened wetlands', http://www.wetlands. org/Aboutwetlandareas/Threatenedwetlandsites/tabid/1125/ language/en-US/Default.aspx; Hunt 2004, p.198,

17. '...excessive enrichment of rivers, lakes, and shallow sea areas, priumarily by nitrate fertilizers washed from the soil by rain, by phosphates from fertilizers, and from nutrients in municipal sewage...', Brimblecombe et al 1998.

18. Cooper and Gilliam 1987 and Cooper et al. 1987.

19. Evans et al 1996.

20. *Ibid.*

21. We are also grateful to Carl Hoffman of the Danish National Environment Research Institute for his clear explanation of the process of nitrate removal by wetlands (as well as shallow bodies of water) in his paper 'Healthy and safe water for people and nature:

Ecological engineering in Denmark', delivered at the EEB Conference 'Crunch time for Europe's water', Brussels, 18 November 2008.

22. IPCC 2008, p.55.
23. Lundqvist et al 2008, p.12.
24. UNDP 2006.
25. IPCC 2008, pp.56–7.
26. IPCC 2008, pp.57–61.
27. IPCC 2008, p.77.
28. Ackerman and Stanton 2006, p.17.
29. Ackerman and Stanton 2006, pp.24–9 contains a review of the various multitrillion dollar estimates of the cost of climate change overall.
30. IPCC 2008, pp.78–9.
31. Research by the Hadley Centre has shown that in higher latitudes rainfall will increase. Hadley Centre 2005, p.34.
32. Lynas 2007, pp. 69–70.
33. IPCC 2008, p.125.
34. IPCC 2008, p.126.
35. Pearce 2006, p.273.
36. UNDP 2006, p.26; Andreu et al 2006, pp.26–46, pp.196–8.
37. Pearce 2006, p.274.
38. UNDP 2006, pp.26–7; Bouguerra 2006, pp.51–2; Andreu et al 2006, pp.198–200; Pearce 2006, pp.291–3.
39. Pearce 2006, pp.287–9.
40. UNDP 2006, pp.26–7.
41. Pearce 2006, pp.281–9.
42. Pearce 2006, pp.297–311; Bouguerra 2006, p.54.
43. IPCC 2008, p.127.
44. WHO/UNICEF 2000, Section 4.4 'Accounting for water loss'; the UK figure is cited in Burston 2006, p.12.
45. OECD 2004, p.3.
46. Hunt 2004, pp.97–100.
47. Hunt 2004, pp.67–8.
48. Stan Cox, 'The Folly of Turning Water Into Fuel', *AlterNet*, 22 March 2008, http://www.alternet.org/water/79957/
49. George Monbiot, 'The Freshwater Boom is Over, Our Rivers Are Starting to Run Dry', *The Guardian*, 10 October 2006.
50. IPCC 2008, p.43.
51. UN 2003, p.13.
52. South Australia Murray Irrigators (SAMI), 'Over-allocation now a global phenomenon: A special report about river management worldwide', *SAMI News*, July 2008, http://www.samurrayirrigators.org.au/sami_news.asp

53. IPCC 2008, p.81.
54. J.W. Maurits la Rivière, 'Threats to the world's water', *Scientific American*, No. 261(3), Sept 1989, pp.80–94.
55. IPCC 2008, p.82.
56. IPCC 2008, p.84.
57. Falkenmark et al 2001, p.13.
58. Falkenmark et al 2001, pp.19–20.
59. Falkenmark et al 2001, pp.31–2.
60. Hunt 2004, p.51.
61. McCully 2001, pp.141–4, emphasis in original.
62. McCully 2001, p.145.
63. Hunt 2004, pp.207–8.
64. Preface by Davis Runnalls to Steenblik 2007, p.vl.
65. Steenblik 2007, p.1.
66. Steenblik 2007, Executive Summary (pp.1–7).
67. Stan Cox, 'The folly of turning water into fuel', *Alternet*, http://www.alternet.org/story/79957/, 22 March 2008.
68. Paul Taylor, 'Biofuels Down, Energy Saving Up in EU Climate Plan', *Planet Ark Daily News*, 29 July 2008, http://www.planetark.org/dailynewsstory.cfm/newsid/49554/story.htm
69. Valerie Mercer-Blackman, Hossein Samiei and Kevin Cheng, 'Biofuel Demand Pushes Up Food Prices', *IMF Survey Magazine*, 17 October 2007, http://www.imf.org/external/pubs/ft/survey/so/2007/RES1017A.htm
70. Lesley Wroughton, 'Biofuels Major Driver of Food Price Rise – World Bank', *Planet Ark Daily News*, 30 July 2008.
71. Hunt 2004, p.209.
72. Hunt 2004, pp.209–10.
73. Goldemberg 2000, Chapter 6: Jochem Eberhard, 'Energy end-use efficiency', p.174.
74. Hunt 2004, pp.212–14.
75. See Hunt 2004, pp.215–22 for a description of renewable energy sources and an assessment of their potential.

5   The European Union Within its Borders: Why Privatisation? The Ideology Behind the Theft of Public Property

1. Traditional rhyme, south-eastern England.
2. Green 2000.
3. The founding texts of what became neo-liberalism are Friedman 1962 and Hayek 1960; see also Hayek 1979 and Weingast 1995. For a thorough critique of neo-liberalism and the theoretical perspective

underpinning this section of the book and influencing the whole of it, see Gill 1995; Gill 1998; van Apeldoorn 2002.

4. Interview conducted by author (McGiffen), Greenwich University, 21 October 2008.
5. Interview conducted by author (McGiffen), Greenwich University, 21 October 2008.
6. Interview conducted by author (McGiffen), Greenwich University, 21 October 2008 .
7. WRc/Ecologic 2002.
8. Hall 2003a, p.2.
9. *Ibid.*
10. Hall 2003a, p.3.
11. Marx and Engels 1875, Part 1.
12. Hall 2003a, p.3.
13. Interview conducted by author (McGiffen), Greenwich University, 21 October 2008.
14. CEC 2003.
15. CEC 2003, p.13.
16. Hall 2003b, p.2.
17. CEC 2003, p.13.
18. Hall 2003b, p.2.
19. CEC 2003, p.13.
20. Hall 2003b, p.2.
21. 'Public procurement: infringement proceedings against Italy and Germany concerning waste management services', European Commission Press Release, 3 April 2008.
22. Corporate Europe Observatory Press Release, 'Moratorium needed to protect public water from misguided liberalisation push', 10 October 2008.
23. CEC–DG–REGI 2003, p.7.
24. Water Time Case Studies, http://www.watertime.net/wt_cs_cit_ncr.asp
25. CEC–DG–REGI 2003, p.52.
26. Hall 2003b, p.4.
27. Hall 2003a.
28. Lobina and Hall 2001, p.4.
29. *Ibid.*
30. 'The Great Water Robbery', *Daily Mail*, 11 July 1994.
31. House of Commons Select Committee on Environmental Audit, Minutes of Evidence taken before the Environment Audit Committee, 28 February 2001, Para 208.
32. Lobina and Hall 2001, pp.13–15.
33. Lobina and Hall 2001, p.16.

34. Lobina and Hall 2001, pp.16–18.
35. Lobina and Hall 2001, pp.18–19.
36. Lobina and Hall 2001, p.19.
37. *Ibid.*
38. Lobina 2001, pp.20–1.
39. Helen Jackson (Labour MP for Sheffield, Hillsborough) speaking in the House of Commons, 1 April 1996, http://hansard.millbanksystems.com/commons/1996/apr/01/drought-yorkshire
40. Lobina and Hall 2001, p.22; Bouchier 1998, Ch.3, 'Lessons learnt from outbreaks of waterborne cryptosporidiosis', 3.1 Review of incidents; 'Cryptosporidiosis', Wikipedia article last updated 2 September 2008, http://en.wikipedia.org/wiki/Cryptosporidiosis
41. Hall 2008, p.4.
42. Green 2005.
43. Orwin 1999.
44. *Ibid.*
45. Chong et al 2006, p.20.
46. Formed by the merger of Gaz de France in July 2008.
47. For a more detailed account of the recent history of French water transnationals, see Holland 2005, Chapter 3.
48. Godoy 2003.
49. Lobina 2000, p.3.
50. Godoy 2003.
51. Lobina 2000, p.4.
52. *Ibid.*
53. Godoy 2003.
54. *Ibid.*
55. Hall and Lobina 2007, p.5.
56. CdCF 1997; PSIRU 2000.
57. Water Remunipalisation Tracker, http://www.remunicipalisation.org/
58. Vinnari and Hukka 2007.
59. Lariola and Danielsson 1998, p.16.
60. Lariola and Danielsson 1998, p.24.
61. See Lariola 2000.
62. Lobina and Hall 2006, pp.13–14.
63. Mohajeri 2003, p.322.
64. Gustafsson 2004.
65. *Ibid.*
66. Gustafsson 2004.
67. See Hukka and Seppälä 2005.
68. Hall 2003a, p.6.
69. Hall and Lobina 2007, pp.2–3.

70. De la Motte 2004.
71. Mohajeri 2003, p.320.
72. Belgium 2006.
73. Hall and Lobina 2007, pp.2–3.
74. Interview conducted by author (McGiffen), Greenwich University, 21 October 2008.
75. Hall 2008, p.30.

6   European Union Within its Borders, Part 2: The Water Framework Directive

1. Interview with the authors, Brussels, June 2008. The interview was conducted in Dutch and Hans Blokland's answers were translated by the authors.
2. CEC 2007c.
3. These and all subsequent quotes from Pieter de Pous (unless otherwise stated) are from an interveiw conducted by the authors in Brussels in June 2008.
4. The Conference was organised in Brussels by the EEB and WWF, 18 November 2008. The speakers referred to were, respectively, Thomas Uhlendahl, Felipe Fuentelsaz and Greig Stuart.
5. These and all quotes from Sergiy Moroz (unless otherwise stated) are from an interview conducted by the authors in Brussels in June 2008.
6. WISE, undated, p.1.
7. OECD 2001, p9.
8. OECD 2004, pp.4–5.
9. CEC 2007c.
10. Nonpoint source pollution is water pollution originating from a number of diffuse sources – such as pollution runoff from farmland. Point source pollution is where the polluting substance enters the water from a single identifiable source, such as a pipe discharging from a factory.
11. Kuneman 2002.
12. Lanz and Scheuer 2001, p.8.
13. Tanja A. Börzel, 'Non-compliance in the European Union. Pathology or statistical artefact', *Journal of European Public Policy*. Vol.8, n.5, 2001, p.818, cited in Carafa 2008.
14. CEC 2004, p.6.
15. CEC 2008, p.6.
16. *Blueprint for Water*, http://www.wcl.org.uk/blueprintforwater.asp
17. OECD 2004, p.3.

18. Caroline Falk, now Assistant General Secretary of the European United Left/Nordic Green Left parliamentary group, but at the time an environmental policy adviser to the group's MEPs, handled the WFD dossier and confirms this view, which is clearly consensual on the left and amongst Greens and environmental NGOs. 'The water legislation was put forward at a time when the Commission, the Council and the Parliament were still making things a little greener, especially when we worked together,' she told the authors (interview, June 2008).
19. Lanz and Scheuer 2001, p.10.
20. EEB/WWF 2006.
21. EEB/WWF 2008, p.5.
22. CEC 2007a, p.4.
23. CEC 2007a, pp.5–6.
24. CEC 2007a, p.6.
25. CEC 2007a, pp.7–8.
26. EEB/WWF 2008, pp.5–6.
27. CEC 2007b, p.16.
28. CEC 2007c.
29. Gleick 1997, p.25.
30. Gleick 1997, p.30.
31. *Ibid.*
32. Miller 2006, p.50.
33. Fuentelsaz 2008.
34. WWF 2004b, p.33.
35. Presentation by Richard Seeber, MEP, on his report 'Addressing the challenge of water scarcity and droughts in the European Union', European Parliament Committee on the Environment, Public Health and Food Safety, Meeting of 26 May 2008 (through an interpreter).
36. CEC 2007a.
37. EEB 2007.
38. CEC 2007a.
39. *Ibid.*
40. EEA 2003, pp.5–10.
41. EEB 2007a.
42. *Ibid.*
43. Jochen Schanze, 'Long-term planning of flood risk management', in Moren-Abat et al 2007, p.87.
44. Cited in Abigail Howells, 'Green Paper on Climate Change and Adaptation Measures', in Moren-Abat et al 2007, p.102. (This is an article explaining the Green Paper, rather than the Green Paper itself; Howells is an employee of the European Commission's Directorate-general for Employment.)

45. EEB 2007a.
46. 'European Floods Legislation Attempts to Manage Risks', *Environment News Service* (ENS), 28 April 2007, http://www.ens-newswire.com/ens/apr2007/2007-04-28-03.asp
47. 'Agricultural Water Saving Obligations Delay', *European Water News*, 28 July 2008.

7 The European Union Beyond its Borders

1. CEO 2003a, p.1.
2. Interview with Olivier Hoedeman, Amsterdam, June 2008 (Hoedeman Interview 2008).
3. *Ibid.*
4. Hoedeman 2007.
5. *Ibid.*
6. *Ibid.*
7. Hoedeman Interview 2008.
8. Hoedeman 2007.
9. Hoedeman Interview 2008.
10. Hall, Lobina and de la Motte 2005, pp.287–8.
11. Hall, Lobina and de la Motte 2005, Table 1, pp.289–91.
12. Hoedeman Interview 2008.
13. *Ibid.*
14. See for example Hall, Bayliss and Lobina 2001.
15. Hall 2008, p.11.
16. Hall 2008, p.13.
17. UNDP 2006.
18. Email from Emanuele Lobina to Olivier Hoedeman, 3 April 2008.
19. Lobina and Hall 1999, pp.6–8.
20. Hall 1999, p.6.
21. Public Citizen 2003.
22. Bayliss 2001, pp.7–9, p.12.
23. See Chapter 5.
24. PSIRU 2000a, pp.2–3.
25. Hall 2001a, p.3.
26. Hall 2001a, p.6.
27. Hall, Bayliss and Lobina 2002, p.36.
28. Winpenny 2003, p.14.
29. See WSSCC information brochure, http://www.wsscc.org/fileadmin/files/pdf/publication/WSSCC_Information_Brochure_2008.pdf
30. 'Financing: Camdessus report lacks pro-poor focus', IRC International Water and Sanitation Centre website, 24 March 2003, http://www.irc.nl/page/2722

31. *Europa World*, 25 April 2003, http://www.europaworld.org/week126/commissionproposes25403.htm
32. CEO 2003.
33. *Ibid.*
34. Official summary of the EU GATS requests, July 2002, quoted in CEO 2003a.
35. EUWI 2008, p.3.
36. Tearfund/Water Aid 2006, p.1.
37. Tearfund/Water Aid 2006, p.3.
38. Tearfund/Water Aid 2006, p.4.
39. Communication of 26 January 2004 from the Commission to the Council and the European Parliament on the future development of the EU Water Initiative and the modalities for the establishment of a Water Facility for ACP countries [COM(2004) 43 final].
40. Gruppo Soges 2008, p.8.
41. CEO 2007, p.3.
42. See Olivier Hoedeman and Orsan Senalp, 'Turkey Plans to Sell Rivers and Lakes to Corporations', Alternet, 23 April 2008, http://www.alternet.org/story/83304/
43. Hall and Lobina 2006, pp.11–12.
44. Hall and Lobina 2006, p.12.
45. Hall and Lobina 2006, p.51.
46. Hall and Lobina 2006, p.36.
47. Hall and Lobina 2006, p.52.
48. CEC 2003b, paras 4 and 62.
49. CEO 2007, pp.2–3.
50. EuropeAid 2007, p.14.
51. CEO 2007, p.4.
52. BizClim press release of 27 November 2006, quoted in CEO 2007.
53. CEO 2007, p.5.
54. UNDP 2006, pp.9 and 15.
55. Resolution of the European Parliament of 11 March 2004 on the strategy for the internal market priorities 2003–2006, Recital 5.
56. European Parliament resolution on the Fourth World Water Forum in Mexico City (16–22 March 2006).
57. UNDP 2006, p.18.
58. UNDP 2006, p.19.

8 A Better Water Policy is Possible

1. The whole resolution is available at http://www.europarl.europa.eu/sides/getDoc.do?pubRef=-//EP//TEXT+TA+P6-TA-2006-0087+0+DOC+XML+V0//EN&language=EN

2. Gleick 1999.
3. It is undoubtedly about time that this was changed to 'his or her', but it hasn't been, possibly at least partly because many languages do not distinguish the gender of the subject in possessive pronouns.
4. *Universal Declaration of Human Rights* (United Nations, 1948), http://www.udhr.org/udhr/default.htm
5. Salman and Bradlow 2006, p.3.
6. Quoted in Gleick 1999, on-line version, p.7.
7. Article 8 (1), http://www.unhchr.ch/html/menu3/b/74.htm
8. Article 24 (2)c, http://www.unhchr.ch/html/menu3/b/k2crc.htm
9. UNEP 1992, Article 18.2.
10. UNEP 1992, Articles 18.8 and 18.9.
11. UNEP 1992, Article 18.19.
12. UNEP 1992, Article 18.16.
13. Hall and Lobina 2006a, p.8.
14. *Ibid.*
15. See Food and WaterWatch website, http://www.foodandwaterwatch.org/, for updated reports on these conflicts, as well as those in the US and other developed countries.
16. See the Network's website http://europeanpublicwaternetwork.blogspot.com/ for details of how you can get involved.

INDEX

accountability 137
ACP 208
  Business and Investment Forum
    216
  see also Business Climate
    Facility
afforestation 65, 119
Africa 25, 35, 80, 205, 208, 216
  East 101, 112
  Horn of 35, 44
  North 25, 44, 111, 210
  Sahel 101
  Savannah 49
  Southern 101, 112
  Sub-Saharan 25, 26, 44, 114,
    212, 213
  West 207, 225
Africa–Caribbean–Pacific *see*
    ACP
African Business Roundtable 216
agriculture 49, 51, 66, 103–4,
    107, 112, 116, 167, 176,
    186–7, 192, 193–4
  and crop rotation 110
  cross compliance 193
  and fertilisers 110, 112, 115,
    173
  intensive 115
  modern monocultural systems
    37
  and pesticides 110, 112, 173
  traditional systems 36
  see also Common Agricultural
    Policy; drought; flood;
    irrigation
Agu, Augustine 43

aid *see* development aid
air conditioning 103
air quality 117
algal blooms 175
Algeria 45
Alston, Margaret 40, 42
Alternative Conjunctive Use
    (ACU) 30
America, Latin 101, 200, 205,
    208, *see also* America, South
America, North 24, 99, 200, *see
    also* Canada; Lakes, North
    American Great; United
    States
America, South 75
  Pampas 49
Andreu, Joaquin 20
Andreu and Solera 20
Anglian Water 155
Angoulême, France 149
Annals of Botany 52
Antwerp 67
Aquafed 198
aquifers *see* groundwater
Aral Sea 22
Argentina 31, 204
Ariès, Paul 15, 16
Arizona 53
Asia 32, 35, 141
  East 101, 211
  South 26, 101, 111, 211, 214
  South-East 86
  Steppe 49
Association of British Insurers
    191